The Institute of Biology's
Studies in Biology no. 11

Muscle

by D. R. Wilkie M.B., B.S., M.R.C.P.(London),
M.D.(Yale)
Jodrell Professor of Physiology,
University College, University of London

Edward Arnold (Publishers) Ltd

First published 1968
Reprinted 1969
Reprinted 1970

Boards edition SBN: 7131 2190 4
Paper edition SBN: 7131 2191 2

Printed in Great Britain by
William Clowes and Sons Ltd, London and Beccles

General Preface to the Series

It is no longer possible for one textbook to cover the whole field of Biology and to remain sufficiently up to date. At the same time students at school, and indeed those in their first year at universities, must be contemporary in their biological outlook and know where the most important developments are taking place.

The Biological Education Committee, set up jointly by the Royal Society and the Institute of Biology, is sponsoring, therefore, the production of a series of booklets dealing with limited biological topics in which recent progress has been most rapid and important.

A feature of the series is that the booklets indicate as clearly as possible the methods that have been employed in elucidating the problems with which they deal. There are suggestions for practical work for the student which should form a sound scientific basis for his understanding.

1968

INSTITUTE OF BIOLOGY
41 Queen's Gate
London, S.W.7.

Preface

Muscles are fascinating to work with because they so obviously *do* something. As our knowledge of the mechanism by which they operate is advancing by leaps and bounds, it seems likely that muscle may be the first tissue whose function is completely understood in terms of ordinary physics and chemistry. It is hoped that this book, by creating interest and providing a little information, may help to accelerate that result.

London, 1968

D.R.W.

Contents

Introduction

It is only by the use of our muscles that we are able to act on our environment—to exert forces and to move objects, including ourselves, around the world. The muscles are biological machines which convert chemical energy, derived ultimately from the reaction between food and oxygen, into force and mechanical work; it is the purpose of this book to explain what is now known about the manner in which this machine works. It is useless to attack such a problem from one narrow point of view, and you will find that it is necessary to apply ideas and experimental techniques derived from mechanics, biochemistry, microscopy, molecular biology, electronics and thermodynamics, in order to find out what is going on. It will be assumed that you already have a background knowledge of these subjects and, more important, a genuine interest in them.

Even unicellular animals, such as the amoeba, can move, though they have no specialized 'muscles' that can be identified under the microscope: however, in most multicellular organisms some of the cells have become specialized for this particular form of energy conversion. In the higher animals muscle constitutes a large fraction of the body, roughly 40 per cent in the case of a man. The 'meat' of the body is almost pure muscle, so is the heart; and the intestines and some other viscera such as the uterus and bladder contain a large fraction of muscle cells too. In spite of the apparent differences in muscular activity found throughout the animal kingdom, it nevertheless appears that the essential biochemical mechanism is everywhere the same—the machinery is composed of protein, which can usually be identified as actomyosin (see p. 13); and the fuel is universally adenosine triphosphate, called ATP for short.

In examining different types of muscle we are therefore concerned to find out how this basic mechanism is adapted to suit the different situations that have arisen in the course of evolution. One basic fact about the contractile mechanism is that, in order to exert a force, chemical energy must be continuously expended. In this respect it is unlike a rubber band, for example, which can exert force continually without cost. Secondly, the slower the contraction of the muscle, the more economically is this force maintained: but the mechanical power that can be produced is correspondingly reduced. A compromise must therefore be made to suit each situation. In our arm muscles a high power production is required; this can only be achieved by making the muscles uneconomical at exerting a force, as we soon discover when we try to hold up a heavy suitcase. In contrast, the folded hind limb of many quadrupeds—such as the cat— would collapse if the muscle within it did not exert force continuously; these postural muscles are accordingly found to be slow in action, but more economical. An extreme example of force being maintained for a long period of time is provided by bivalves such as the oyster and mussel. These can hold their shells shut tight for many hours because they have evolved muscles that can maintain tension very economically, yet they have also evolved a special mechanism to switch the tension on and off again reasonably rapidly.

Other variations on the theme are shown by heart muscle, which has a built-in mechanism to maintain rhythmic contraction of the whole muscle, quite independently of any nervous connections; and the fibrillar muscle that actuates the wings of many insects. This does not shorten like an ordinary muscle, but if it is suitably loaded it generates rapid vibrations.

These functional differences are to a large extent expressed as structural differences, which will be discussed in more detail in Chapter 3. Skeletal, cardiac and fibrillar muscle fibres all have a striking pattern of *cross-striation*. Other types of muscle are called *smooth* but it must be clearly realized that the smooth muscles comprise a very heterogenous group. The smooth muscles of vertebrate viscera are composed of small spindle-shaped

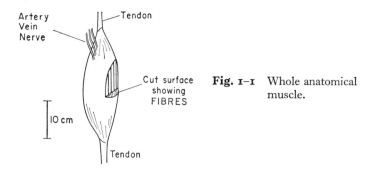

Fig. 1-1   Whole anatomical muscle.

cells with virtually no internal structure, while many invertebrate smooth muscles contain quite highly differentiated systems of protein filaments.

In muscle, the relation between structure and function is very close, so in order to present the subject on a concrete basis, let us first consider the detailed structure of a vertebrate skeletal muscle, as shown in Figs. 1-1 to 1-8. This type of muscle is shown because more research has been done on it than on the other types. We are concerned with eight orders of structure of diminishing size—from gross anatomical down to interatomic dimensions, as indicated roughly by the scales. Note the units used for measuring lengths—centimetres for gross dimensions, μ (pronounced "mu", $1\,\mu = 10^{-4}$ cm) for microscopic ones and Å (pronounced "Ongstroem" or "Angstrom", $1\,\text{Å} = 10^{-4}\,\mu = 10^{-8}$ cm) for ultramicroscopic or molecular ones.

Whole anatomical muscle (Fig. 1-1). As mentioned above, the muscles are the meaty part of our bodies. There are more than 150 different anatomical muscles and almost all of them are attached to the skeleton at both ends, frequently through a strong tendon.

Muscles can only pull—they cannot push—and in order to produce the complex movements of the body the muscles have to act together, but in changing patterns, on the already complex lever system provided by the skeleton, as will be explained in Chapter 7.

The muscle needs an artery and vein so that during exercise it may be abundantly supplied with oxygen, which is required in order to release energy.

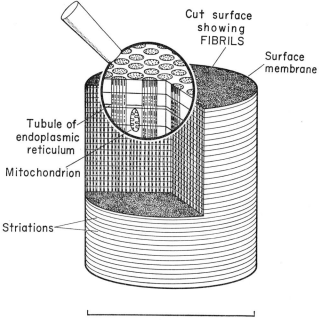

Fig. 1–2 A short segment of a single muscle fibre.

It also needs a nerve supply so that its contraction can be regulated according to the instructions sent down along motor nerve fibres from the central nervous system, that is, the brain and spinal cord. Finally, in order to co-ordinate movement, the central nervous system needs to be informed about the actual length of the muscle and about the tension in its tendons. This information is provided by special sense organs and signalled along sensory nerve fibres.

The muscle has a grainy appearance because it is made up of smaller subunits, the fibres.

The muscle fibre (Fig. 1–2). This is a cylindrical structure which may be many centimetres long. After staining, or when suitably illuminated, it is seen to have regular striations which extend right across inside the fibre, dividing it up into *sarcomeres* which are stacked one on top of the other like coins in a pile. The details of this striation will be dealt with under Fig. 1–3.

Within the fibre can be seen many cylindrical subunits, the *fibrils*. These are the structures that actually contract, and, in between them, in the *sarcoplasm*, are other structures of functional importance. The *mitochondria* are the chemical "factories" in which most of the oxidation of foodstuffs occurs (see Chapter 6). The complicated, ramifying tubules of the *endoplasmic reticulum* appear to be important as part of the mechanism by which contraction is switched on and off, as explained on p. 44. The *surface membrane* is certainly important in this connection too, for

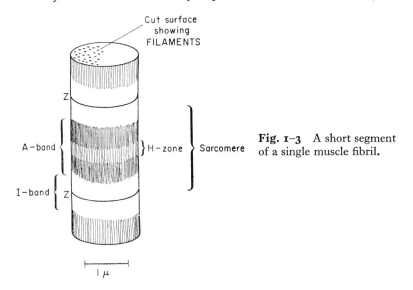

Fig. 1–3   A short segment of a single muscle fibril.

it only allows certain ions to pass through; this leads to the generation of the resting potential (p. 42) and of the action potential (p. 44).

In the living body the process of excitation and contraction is only set going when an action potential arrives down the motor nerve fibre from the central nervous system. The nerve fibre is connected to the muscle fibre at a special junction region called an *end-plate*, whose properties are described on p. 44.

The fibril or myofibril (Fig. 1–3) is a rod of contractile protein which runs from one end of the fibre to the other. In life it is perfectly transparent, but if it is observed through a special microscope that detects differences of refractive index or of polarization, a pattern of cross-striations can be seen. The correct understanding of this pattern has led to great advances in our knowledge about contraction, as explained on p. 19.

The fibril is divided up into segments by thin partitions called Z-lines or Z-discs. These also run from fibril to fibril right across the fibre, thus dividing it into sarcomeres. In the middle of the sarcomere is the A-band of high refractive index, with a less refractile central H-zone. The rest of the sarcomere is occupied by I-band.

The fibril itself is composed of longitudinal fine protein *filaments*.

The protein filaments (Fig. 1–4) are of two types, thick and thin, and they interdigitate with each other as shown in the diagram. The H-zone is that part of the A-band from which thin filaments are absent. Except for a short region in the middle, the thick filaments have projections sticking out to either side. These projections, or cross-bridges, are thought to be the places where the force of contraction is actually developed.

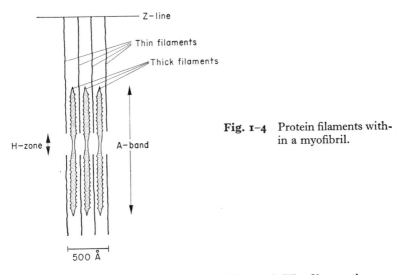

Fig. 1–4 Protein filaments within a myofibril.

The internal structure of a thin filament (Fig. 1–5). The filament is composed of two strings of globular protein units wound round each other to form a two-stranded rope. The structure of the thick filaments is more complex and is described on p. 11.

In order to understand the structure of the globular subunits it is best to start with the chemical structure of a protein.

The chemical structure of a protein (Fig. 1–6). It is formed by the polymerization of amino acids. Each amino acid has the formula

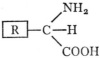

where R — is one of twenty or so different chemical groups that characterize the different amino acids. The —COOH group of one amino acid readily reacts with the —NH₂ group of another, with the elimination

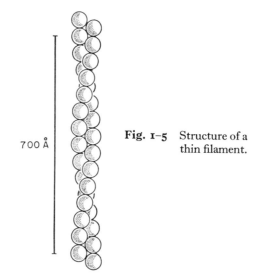

Fig. 1-5 Structure of a thin filament.

of H₂O, and this leads to the structure shown in Fig. 1-6. This has a central spine of —C—C—N—C—C—N— atoms, with the $\boxed{R}$ groups sticking out sideways.

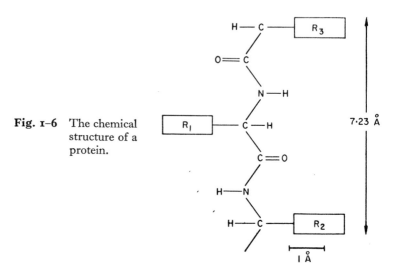

Fig. 1-6 The chemical structure of a protein.

Secondary structure: the α-helix (Fig. 1–7). The central spine readily curls up into a helical (corkscrew-shaped) coil, as shown. The $\boxed{R}$— groups then stick out radially like the bristles on a test-tube brush, so that the whole forms a cylindrical structure.

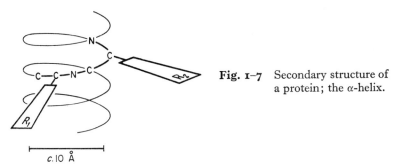

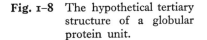

c. 10 Å

Fig. 1–7 Secondary structure of a protein; the α-helix.

Tertiary structure (Fig. 1–8). In order to form the globules shown in Fig. 1–5, the α-helix bends on itself as shown, so that there are relatively straight sections of α-helix with bends between them where the arrangement is less regular. In fact the tertiary structure of a muscle protein has

Fig. 1–8 The hypothetical tertiary structure of a globular protein unit.

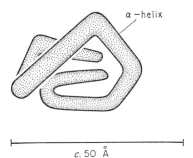

α –helix

c. 50 Å

not yet been determined: Fig. 1–8 is based on what was found in haemoglobin, another globular protein.

1.1 Experimental methods

Anatomical dissection and observation suffices for Fig. 1–1. However, the functional significance of what is seen is not so obvious as you might imagine, as explained on p. 57. Light microscopy would reveal most of Figs. 1–2 and 1–3, but remember that since a living muscle fibre is perfectly transparent, you must use a phase contrast or interference microscope (sensitive to differences of refractive index) or a polarizing microscope (sensitive to optical anisotropy: quartz crystals show anisotropy and so does the A-band, hence its name). With an ordinary microscope, slightly out of focus, a sort of phase-contrast effect makes it possible to visualize striations, but you have to be very careful over identifying them, as the A-band appears light or dark according to the adjustment of the microscope. This has led to much confusion in the past.

A light microscope cannot resolve objects closer than the wavelength of light—say $0.5\,\mu$—so for Figs. 1–3, 1–4 and 1–5 an electron microscope would be used. X-ray diffraction (see p. 21) detects structures that repeat regularly with a period from less than 1 Å up to several hundred Å, so this method has been used to confirm Figs. 1–4 and 1–5 and to establish Figs. 1–7 and 1–8. Finally, the amino acid sequences in Fig. 1–6 must be determined by purely chemical means.

The composition of muscles 2

In this chapter we are concerned simply with the composition of muscle as shown by chemical analysis. The different constituents will be discussed in groups according to their functional significance. In the next chapter we will take up the question of the structural arrangement of the different substances at a microscopic and ultramicroscopic level. Chemical analysis may seem a rather prosaic approach to the question of how muscles actually work—nevertheless, even today, it presents us with some major un-answered questions; certain substances are present in quite high con-centration without having any known function. The most challenging substances are the iminazole base carnosine (in some species replaced by anserine) which has not so far been shown to play an intelligible part in contraction or recovery though it may act as a buffer; and zinc, which is present in almost the same concentration as calcium but has not yet been assigned a functional role in normal contraction.

2.1 Water

Muscle contains about 80 per cent water and it is not there simply to fill up the space: it plays a vital role in contraction because of its unique physico-chemical properties. It cannot be successfully replaced even by the closely related compound deuterium oxide, D_2O. Part of the unique-ness of water arises from the fact that even at room temperature the water molecules are not randomly arranged like the molecules of a perfect gas. On the contrary, because of the polar nature of the individual molecules, water has a marked tendency to form liquid crystals, especially if given an electrically charged ion or a membrane (such as abound in muscle) to act as a nucleus around which water molecules can arrange themselves. The formation and dissolution of such hydration shells are accompanied by quite large changes of energy, just as they are when a strong acid is neutralized by a strong alkali, and these effects may well be vital to the normal functioning of the tissue, though at present there is no decisive evidence on the point.

Under normal conditions about 20 to 25 per cent of the water is present in the interspaces between the fibres. Movement of water in and out of the fibre is regulated by osmotic forces; the water tends to move into regions where the concentration of solute is highest. Thus in weak solu-tions the fibres tend to swell and in strong ones they shrink. Consequently, one of the important considerations in making up a physiological solution (see p. 23) in which the muscle will function normally, is that its osmolarity should be the same as that of the interior of the muscle fibre. There is still a good deal of discussion over the question whether the water in muscle is 'free' to take part in physico-chemical processes or whether part of it is 'bound' to protein so tightly as effectively to separate it into a separate compartment. In support of the idea that part of the water is bound is the fact that not all the water freezes at moderately low temperature, and that an exceedingly dry atmosphere, i.e. a very low water vapour pressure, is needed to dry a muscle completely: moreover, although a muscle fibre behaves as a perfect osmometer when it is placed in solutions of various concentrations, it does so as though part of its water was not participating in these changes. On the other hand, all the water seems able to dissolve salts and urea with the normal depression of vapour pressure, so the situation is still not entirely clear.

2.2 Proteins

These make up most of the solid matter of the muscle; this is, incidentally, the reason why meat is such a valuable food. The contractile machinery itself is made of protein, as mentioned in the Introduction. Much of the protein is firmly attached within the muscle fibre as shown by the fact that it is far more difficult to extract the protein from the muscle in the first place than to keep it in solution later. Tests of extractability and solubility are an important way of distinguishing one type of protein from another, since some proteins can be extracted by salt solutions that are too weak to extract others. The key property of the solution that decides how effective it will be in extracting protein is not really the concentration as such, but the *ionic strength*, which is compounded of the concentration and the valency, as follows: multiply the concentration of each ion, in gram-ions per kilogram solvent, by the square of the valency; add all these terms together and divide by 2.

Thus for 0·01 molal Na_2SO_4,

$$\text{ionic strength} = (0·02 \times 1^2 + 0·01 \times 2^2)/2 = 0·03$$

For monovalent salts such as KCl, the ionic strength is numerically equal to the molality.

Other tests to separate and characterize proteins depend on the mobility of the protein in a strong gravitational field (ultracentrifugation) or an electric field (electrophoresis) or through a column of a suitable absorbant (chromatography). Nowadays some proteins can even be identified by the appearance of their individual molecules in the electron microscope. Tests for enzyme activity or for reaction with specific antibodies can also be extremely useful in special cases.

From the functional point of view, the many proteins in a muscle fibre can be classified into three main groups:

1. Stroma protein. About one fifth of the protein is very insoluble and it seems to function solely as an inert structural element, or skeleton, to hold the rest of the structures in place. Part of this protein is extracellular and can be identified with the collagen and elastin filaments that bind the fibres together and transmit their tension to the tendons. The rest of the stroma protein plays an analogous role within the fibres.

2. *Ordinary cellular proteins.* These are the proteins that are not specifically characteristic of muscle but are also found in other metabolically active cells, and they comprise about another fifth of the total protein. The most interesting of these proteins are the enzymes, more than fifty of which are responsible for guiding the chemical reactions within the cell and thus for keeping the contractile system supplied with energy. Some of the enzymes can be extracted easily into solution but others, notably those concerned in the oxidation of foodstuffs, are bound to the mitochondria.

3. *Special contractile proteins* are the distinguishing feature of muscle. Two types of protein, *myosin* and *actin*, are known to be absolutely essential for contraction: they constitute roughly 35 and 15 per cent of the total protein respectively. A third type of protein, *tropomyosin*, is also present in appreciable amount (about 10 per cent of the total) and it seems to be important in contraction, but its function has not yet been demonstrated as clearly as those of the other two. The most recent evidence suggests that tropomyosin is involved in sensitizing the contractile proteins to calcium. This sensitivity is important in order to switch contraction on and off.

Myosin can be extracted from fresh minced muscle by salt solutions whose ionic strength is about 0·4 to 0·5. However, a crude extract made in this way contains other proteins, notably some actin, which very much affect its properties. Special solutions have been evolved which do extract myosin in relatively uncontaminated form—the usual one is called Hasselbach–Schneider solution, and it consists of 0·47 M KCl, 0·1 M potassium phosphate buffer pH 6·5, and 0·01 M sodium pyrophosphate. In solution the myosin molecules have a very characteristic shape, as shown in Plate 1 (a); in electronmicrographs they look like spermatozoa (though they are very much smaller) with a compact 'head' region and a long 'tail'. It has been known for some time that the myosin molecule can be split into two fragments by brief digestion with trypsin, the enzyme from the pancreas that normally digests proteins. The two particles correspond roughly with the 'head' and 'tail' of the molecule; they are called heavy meromyosin (shortened to HMM, molecular weight about 350,000) and light meromyosin (LMM, molecular weight about 150,000) respectively. These two parts of the myosin molecule play very different roles in the mechanism. The main function of LMM seems to be structural; it consists of a long rod about 20 Å in diameter and perhaps 1000 Å long, which is probably composed of two α-helices coiled round each other like a two-stranded rope. HMM is more complicated and interesting. Its detailed structure is not yet known but is probably globular and it possesses two different chemically active sites. At one of these binding to actin can occur and at the other the hydrolysis of ATP is catalysed. Thus HMM appears to be the very kernel of the contractile machine, for within this small volume (about 40 Å diameter by 300 Å long) there are the means for producing breakdown of the chemical fuel (ATP-ase site) and also for producing a mechanical effect (binding site).

In solutions of high ionic strength, say 0·4, the individual myosin molecules are not attached to each other, but if the ionic strength is lowered by adding water, the myosin molecules aggregate together to form rods that are big enough to be seen in the ordinary light microscope. The way in which this aggregation occurs is illustrated in Plate 1 (b) and (c). In the middle of the rod is a region about 0·2 μ long which is composed solely of tails. Elsewhere the heads stick out from the surface of the rods, which thus bear a very strong resemblance to the thick filaments of myosin found in living muscles (see also Plate 3). At first sight the orderliness of the organization of a striated muscle, extending down as it does to molecular dimensions, seems to demand an extraordinarily elaborate system of control during the development of the cell. However, the experiments just described show that the orderliness arises quite naturally and demands no more specific direction (and no less) than does the growth of a crystal.

Actin is far more difficult to extract from the muscle, probably because it is structurally attached to the Z-membrane (see Plate 4). However, it can be dissolved by 0·6 M potassium iodide solution and it then remains in solution even after all the salts have been removed, for example by dialysis; that is, suspending the protein solution in a bag whose walls permit the small salt ions to diffuse away, but not the large protein molecules. Actin is then found to consist of roughly spherical molecules about 55 Å in diameter, with a molecular weight of 60,000; in this form it is called globular (or G-)actin. Unlike HMM, actin does not possess ATP-ase activity; but each globular molecule contains one molecule of ATP very firmly bound to it.

If the ionic strength is raised by adding salts, the G-actin molecules polymerize, that is, combine together to form long chains, in the form of a two-stranded rope that was shown in Fig. 1–5: the polymerized form is called F-(for fibrous)actin. The G-actin molecules can only stick together if at the same time their bound ATP becomes hydrolysed to ADP, thus liberating energy:

$$\text{G-actin-ATP} \longrightarrow \text{F-actin-ADP} + \text{inorganic phosphate}$$

Just as in the case of myosin, the filaments that form sponteneously in solution bear a very strong resemblance to those that are found in living muscle; and once again there appears to be a connection between ATP splitting and a mechanical change in the protein.

Tropomyosin is the name given to two somewhat different proteins that are intimately concerned with the contractile mechanism even though their precise function is not certain at present.

Tropomyosin A (TMA, sometimes called paramyosin) is found assoc-iated with myosin in the thick filaments that are characteristic of the muscles of some molluscs. It has no ATP-ase activity and its function may be largely structural: its own structure resembles that of the LMM in the tail of the myosin molecule, being composed of two α-helices twisted together.

Tropomyosin B (TMB) has a similar structure and is thought to give mechanical strength to the Z-disc. However, it may also play a more active role, since a protein that closely resembles TMB has been shown to combine with actin and to have an important effect on the reactions of the con-tractile system to calcium. Sensitivity to calcium is essential for the mechanism that switches the contractile process on and off, as explained on p. 44, and this senstitivity is not shown by pure actin and myosin.

When solutions of actin and myosin are mixed, the two proteins combine to form *actomyosin* and the solution becomes very viscous. The myosin retains its ATP-ase activity, and if ATP is added, together with the co-factors calcium and magnesium, active splitting of ATP occurs. At the same time, the protein becomes precipitated and shrinks down actively to form a dense plug. The significance of this 'superprecipitation' was appreciated in the early 1940's by Albert Szent-Györgyi, who correctly predicted that if the actomyosin molecules were orientated in line instead of lying at random, the result would be shortening and tension develop-ment such as are shown by living muscles.

On the other hand, if the ATP is added under conditions where no splitting of it can occur, for example if there are insufficient calcium ions ($< 10^{-7}$M) to activate the ATP-ase, then actomyosin dissociates again into its component proteins and the viscosity of the solution falls.

These reactions in solution have their close counterpart in the living muscle, as shown in Table 1.

Table 1   Relations between actin (A), myosin (M) and ATP.

	In solution	*In living muscle*
A and M present, no ATP	AM formed, solution becomes very viscous	Rigor: muscle stiff and inextensible
A and M present, ATP present and being split	Superprecipitation	Contraction
A and M present, ATP present but not being split	AM dissociates into A+M, viscosity falls	Relaxation

Table 1 shows why it is that after death the muscles pass from the state of relaxation to one of rigor mortis: their ATP level falls and is not restored by metabolic processes. Incidentally, it is well known in medicolegal work that rigor mortis comes on much more rapidly in muscles that were exercising vigorously just before death.

2.3 Substances for energy storage

The proteins make up the machinery of the muscle cell. It is a machine of enzymes which guide a complex system of chemical reactions so as to produce ATP, and of contractile proteins which break the ATP down and transform part of its energy into mechanical force and work. ATP is the most important store of energy in the sense that it is the *only* substance that the contractile proteins can use directly. Exactly what it is that is so special about the ATP molecule remains one of the fundamental mysteries of biology; but there is no doubt that ATP is vital for energy transformation in such diverse situations as muscular contraction, active transport and bioluminescence. It also acts as an energy store during photosynthesis and in the ordinary biochemical synthesis of many important compounds. Incidentally, measurement of the light emitted by an extract of firefly tails when ATP is added to them is a common and very sensitive way of estimating ATP.

In spite of its great importance, there is not much ATP in the muscle—only about 3 mM/kg or enough for 8 brief contractions (the figures in this section refer to frog muscle). Since a living muscle can obviously perform much more than 8 twitches, it follows that the ATP that is broken down during activity must be very rapidly restored, using energy from other stores. The store closest to hand is of phosphocreatine—about 20 mM/kg, enough for nearly 100 contractions—and this energy is available while contraction is actually continuing. This store must finally be replenished from the much slower processes of recovery which draw their energy from the oxidation of a carbohydrate, glycogen, which is stored in the muscle in the form of granules. An old name for glycogen—animal starch—is quite a good one; like starch, glycogen consists of hexose units polymerized together. This is a very large store, equivalent to 100 mM of hexose per kg of muscle, which would yield on oxidation enough energy for 10,000 to 20,000 twitches. If the supply of oxygen is insufficient, as may occur during moderate exercise, energy may still be obtained from glycogen by hydrolysing it to lactic acid; the total energy is now less, enough for only about 600 twitches. The significance of these different energy stores in exercise will be discussed further on p. 60.

2.4 Inorganic ions

These play a number of important roles in controlling the contractile process. The concentration of each, in gram-ion/kg (as found in frog muscle) will be shown in brackets after each ion is first mentioned. The importance of magnesium (10) and of calcium (4) in controlling actomysin

ATP-ase has been mentioned before. The specially important part played by calcium will be dealt with on p. 45. Potassium (90) and sodium (15) are exceedingly important in setting up the differences of electrical potential upon which the propagation of action potentials depends—see p. 41. These action potentials are the means by which the central nervous system gives the muscle fibres the order to contract.

The structure and ultrastructure of muscle

Although the fundamental biochemical mechanism of contraction involving actin, myosin and ATP seems to be the same in all muscles, the arrangement of the mechanism varies from one situation to another. All muscles are composed of *fibres* which may be single cells or may be syncytia. The classification of the different types of muscle fibre is based primarily on the presence or absence of regular cross-striations, which can be detected by ordinary light microscopy, along the length of the fibres. In vertebrates both the cardiac and the skeletal muscles are striated; and unstriated (smooth) muscles are used for internal control of gut, blood vessels and the viscera. In other phyla, too, striated muscles are usually found in situations where fast action is required, e.g. in the wing muscles of insects, but the correlation between structure and speed is not perfect and some smooth muscles, e.g. the muscles that operate the nictitating membrane of the cat, have quite a fast action.

3.1 Unstriated, smooth or plain muscle

In vertebrates smooth muscles are composed of fusiform cells, each with a single nucleus, as shown in Fig. 3–1; a faint longitudinal striation can be

20–50μ

Fig. 3–1 Diagram of a visceral smooth muscle cell as seen in longitudinal section. Note the spindle-shaped cell with one nucleus and the absence of cross-striations.

made out. These muscle cells contain actin and myosin of normal type, yet in sections prepared for the electron microscope no clear-cut system of protein filaments has yet been demonstrated, so we still do not understand the mechanism of contraction.

In some smooth muscles from other phyla, quite clear systems of filaments have been demonstrated. For example, the muscle that retracts the byssus of the common mussel contains normal-looking actin filaments arranged around large 'paramyosin' filaments, which in turn seem to be composed of tropomyosin A with a small admixture of myosin.

3.2 Striated muscle

The high degree of organization of a striated muscle has been indicated in Chapter 1 and Figs. 1–1 to 1–8. However, it is important to realize that the characteristic striations can only be seen under special circumstances. Through an ordinary microscope, correctly focused, unstained living fibres merely appear as perfectly transparent tubes without any internal structure. It is thus quite mistaken to refer to the A- and I-discs as 'dark' and 'light'; they do not differ at all in their power to absorb light. However, the following differences between the discs can be exploited to make them *appear* dark and light:

1. The A-disc is stained selectively by iron haematoxylin so this can be used for the ordinary histological examination of fixed material.

2. The A-disc is birefringent or anisotropic. In anisotropic materials (of which the best known is calcite) the refractive index depends on the angle of incidence and the state of polarization of the incident light. This is different from what is found in ordinary isotropic materials such as glass, or the I-disc, and the difference can be exploited in a special polarizing microscope to make the A- and I-discs visible, though which one of them appears dark merely depends on the setting of the instrument. Anisotropy is a sign that the structure in question is highly ordered right down to molecular dimensions. The electronmicrograph shown in Plate 3 confirms that this is true of the A-disc.

3. Quite apart from anisotropy, the A-discs have a higher refractive index than the I-discs because they contain a higher concentration of protein. Rays of light are therefore slowed down more if they pass through an A-disc than if they pass through an I-disc. The resulting differences in phase can be converted into a pattern of differences in intensity by arranging that the emergent light interferes with a reference beam, either locally, as in a *phase contrast* microscope, or over the whole field as in an *interference* microscope. Once again, the appearance depends entirely on the setting of the instrument.

4. Even using an ordinary microscope, a sort of phase-contrast effect can be produced by having the microscope slightly out of focus. If it is below proper focus, the A-disc will appear darker; above focus, the converse. This effect was well understood by early microscopists, who could not regard their instruments merely as something bought 'off the shelf'. Later workers were not so careful. The result, as related by A. F. HUXLEY (1957) was a half-century of confusion.

In all types of striated muscle the fibre is subdivided by partitions, the Z-discs, that run right across the fibre, into short segments called sarcomeres which are the functional units of the contractile system. As shown by the labelling on Fig. 3–2, each sarcomere thus contains a whole A-disc and two half I-discs. The whole structure becomes far more intelligible once the ultrastructure is revealed, see Plate 3.

3.2.1 Non-cardiac striated muscle

This is more commonly known as skeletal or voluntary muscle. The second name is a bad one as the very ideas expressed by the words 'voluntary' or 'under the control of the will' are completely meaningless except where they refer to man. There is no possible way of deciding whether a particular movement performed by a frog was 'willed' or not. The fact is that vertebrates use striated muscle both for their predictable and unhesitating (='reflex') and for their unpredictable (='voluntary') movements.

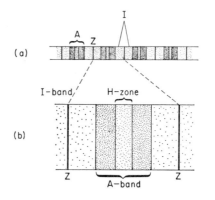

Fig. 3-2 (a) A single myofibril as seen in an interference microscope. (b) Diagram of a single sarcomere from the fibril in (a). (After STARLING and LOVATT EVANS *Principles of Human Physiology.* 13th ed. 1962. Eds. DAVSON and EGGLETON. Churchill, London. Fig. V.3.).

A skeletal fibre such as is shown in Fig. 1-2 might be many centimetres long and contain thousands of sarcomeres. In a resting muscle the sarcomere is about $2\frac{1}{2}$ µ long. This fact may be used to set a scale for Plate 3.

3.2.2 Changes in the pattern of cross-striation related to changes in fibre length

When a whole muscle is stretched, the individual sarcomeres become stretched proportionately. This can be shown in a variety of ways, the most direct of which is to examine sections of the muscle by light or electron microscopy. However, it is difficult to eliminate artifacts produced by the technique of fixation. Hence the value of a different technique, involving the diffraction of light, which can be applied to the whole living muscle. This technique will be described in detail, partly because it requires only simple apparatus, and partly to introduce the principles of the method, which apply equally to x-ray diffraction (p. 21).

Because of its striations, muscle behaves like a diffraction grating with lines spaced apart at a regular interval equal to the sarcomere length, AC=S, as in Fig. 3-3 (a). Each Z-line can be regarded as a new source, scattering light in all directions. Along some directions the scattered waves will all be in step, so the waves reinforce one another and the light is bright.

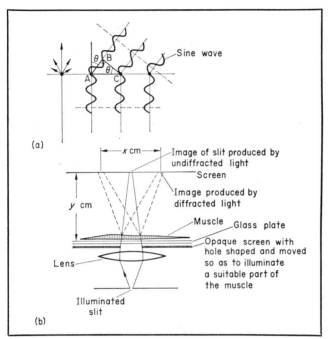

Fig. 3-3 (a) Theory. AC=S, the sarcomere spacing (μ); AB=λ, the wavelength of light (μ). **(b)** Practical arrangement.

The condition for the first bright fringe is that the path difference AB should be one full wavelength, AB=λ, as shown. Then from triangle ABC,

$$\lambda/S = \sin \theta$$

Thus if the wavelength λ is known and the angle θ can be measured experimentally, S can be easily calculated. If S is large compared with λ, other orders of spectra will be seen at larger values of θ corresponding to $2 \times \lambda$, $3 \times \lambda$, etc. One practical arrangement for measuring θ is shown in Fig. 3.3 (b). A slit illuminated by white light has its image thrown by a lens on to a screen. If a striated muscle is interposed and the aperture of the lens reduced so that all the light is obliged to pass through the muscle, rainbow-hued fringes will be seen to either side of the central image of the slit. The yellow fringes (λ=0·6 μ) are the most prominent. If the two yellow fringes are x cm apart and the screen is y cm from the muscle, then

$$\tan \theta = x/2y \simeq \lambda/S$$

therefore $$S \simeq 2y\lambda/x \text{ (microns)}$$

The muscle must be thin if sharp fringes are to be obtained: the sartorius from a small frog is quite suitable When the muscle is stretched, the fringes move closer together, showing that the sarcomere spacing has indeed been increased by stretch.

Of course, this technique can only give information about changes in the sarcomere as a whole. More detail is given by direct microscopy either of living single muscle fibres (A. F. HUXLEY and NIEDERGERKE, 1954) or of individual myofibrils (HANSON and H. E. HUXLEY, 1954). These studies have shown that the A-disc stays practically constant in length; all the variation occurs in the I-discs. The reason for this, which becomes quite apparent when we examine electronmicrographs of muscle (Plate 3), is that the A-disc is composed of thick filaments of myosin held together in a hexagonal pattern (see Plate 4). Thin filaments of actin run from the Z-disc in between the myosin filaments. Both types of filament remain of fixed length; thus when the sarcomere shortens, the area of overlap increases, and vice versa. The thick filaments are 1·5 to 1·6 μ long in a wide variety of vertebrates, hence the A-disc is also 1·5 to 1·6 μ long. The thin filaments vary more from one species to another, ranging from 2·05 to 2·6 μ measured from tip to tip and thus including the Z-region.

Several other interesting facts emerge from Plates 3 and 4.

1. The H-zone is simply the region of the A-disc into which the thin filaments have not penetrated. When the muscle is stretched, the H-zone therefore becomes longer.

2. According to the exact angle at which the section was cut, either one thin filament (a) or two (b) are seen between the thick ones.

3. Cross-bridges can be seen between the thick and thin filaments. It is not obvious from Plate 3, but these do in fact stick out from the thick filaments no matter whether a thin filament is present or not. The cross-bridges are almost certainly the enzymatically active HMM molecules described on p. 11.

4. In the middle of the thick filament is a region about 0·2 μ long which does not have any cross-bridges. The reason for this (see Plate 1 (b), (c)) is that this part of the filament is made up entirely of LMM 'tails'. In the very middle of the A-disc the filaments are thickened, unlike the artificial myosin filament shown in Plate 1 (c), giving rise to a central M-disc, where cross-struts join the myosin filaments together.

The structure of the fibrils from other types of striated muscle, such as cardiac muscle or insect fibrillar muscle differs only in minor respects from the description just given.

Plate 1

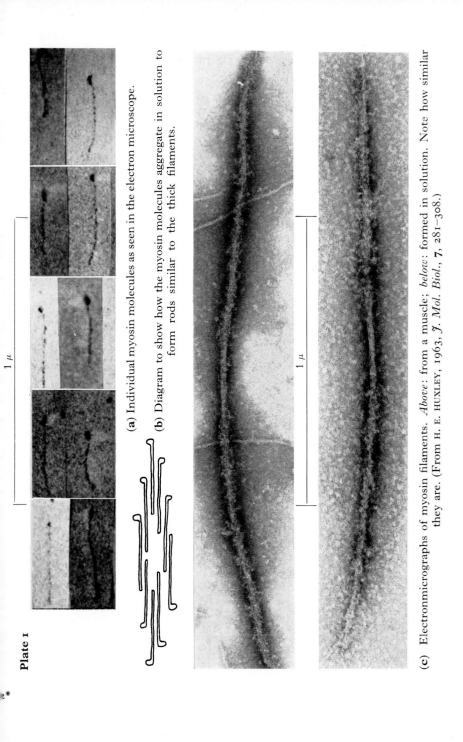

(a) Individual myosin molecules as seen in the electron microscope.

(b) Diagram to show how the myosin molecules aggregate in solution to form rods similar to the thick filaments.

(c) Electronmicrographs of myosin filaments. *Above:* from a muscle; *below:* formed in solution. Note how similar they are. (From H. E. HUXLEY, 1963, *J. Mol. Biol.,* **7**, 281–308.)

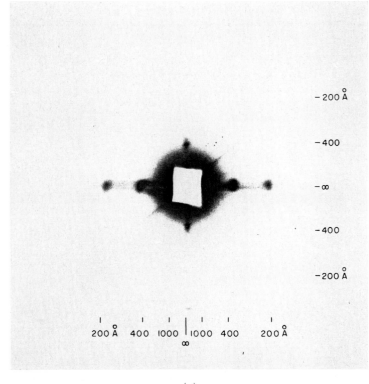

$-200\ \overset{\circ}{A}$

-400

$-\infty$

-400

$-200\ \overset{\circ}{A}$

$200\ \overset{\circ}{A}$	400	1000	∞	1000	400	$200\ \overset{\circ}{A}$

(a)

Plate 2 X-ray diffraction pictures of living muscle at rest. The longitudinal axis of the muscle is vertical. The white rectangle arises from the lead stop that absorbs the direct, undiffracted beam. The diagonal line is an artifact, arising from the system of monochromators. The scales show the spacing on the specimen that would give a first-order spectrum in the position indicated. The scale is a reciprocal one for reasons that will be clear

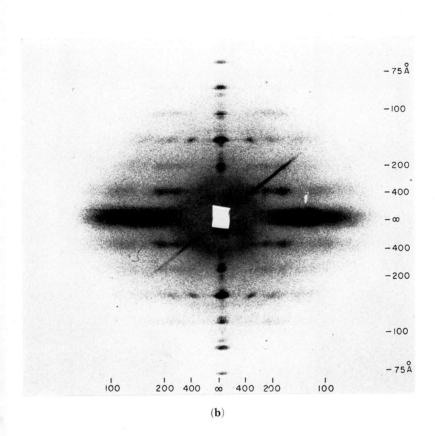

−75 Å
−100
−200
−400
−∞
−400
−200
−100
−75 Å

100   200 400 ∞ 400 200   100

(b)

from Fig. 3–3. Unfortunately, it is impossible to show all the spacings in one photograph, because they need different exposures. (a) has been exposed so as to show the transverse spacing between thick filaments; (b) shows longitudinal dots arising from the myosin filaments, and horizontal lines from the crossbridges. The 59 Å reflection from actin cannot be seen. (Photographs by courtesy of H. E. HUXLEY.)

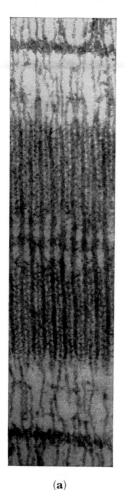

(a)

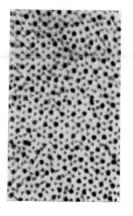

(b)

Plate 3 (**a** and **c**) Longitudinal and (**b**) transverse sections through striated muscle (see Plate 4). The transverse section passes through a region of the A-band where both thick and thin filaments are present. (Electronmicrographs by H. E. HUXLEY from STARLING and LOVATT EVANS, 1962, *Principles of Human Physiology* (13th ed.). Eds. DAVSON and EGGLETON. Churchill, London. Fig. V.9.)

(c)

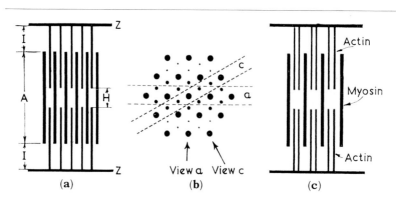

Plate 4 Diagrams to clarify the electronmicrographs shown in Plate 3. (**a**) and (**c**) show longitudinal sections made in the radial planes indicated in (**b**) and accordingly show single or double interdigitation. (From STARLING and LOVATT EVANS, 1962, *Principles of Human Physiology* (13th ed.). Eds. DAVSON and EGGLETON. Churchill, London. Fig. V.8.)

3.3   The sliding filament theory of contraction

This theory was evolved independently and more or less simultaneously by A. F. Huxley and H. E. Huxley round about 1950, and it is now accepted almost universally as a result of the evidence that has accumulated since that time. According to this theory, the force of contraction is developed by the cross-bridges in the overlap region and active shortening is caused by movement of the cross-bridges, which causes one filament to slide or creep over the other. Now the cross-bridges are spaced roughly 450 Å apart, that is, only about 5 per cent of the length of the half-sarcomere. However, both skeletal and cardiac muscle can shorten actively by about 30 per cent so each individual cross-bridge must detach itself from one site on the actin and reattach itself to another site further along, and repeat the process five or six times, with an action similar to that of a man pulling in a rope hand-over-hand.

In some insect flight muscles, on the other hand, the maximum length change is only about 5 per cent and this might well be achieved without making and breaking the links.

Exactly how the minute cross-bridges achieve their complicated action is the central problem in muscle physiology at the present moment—and they are very inaccessible to direct experiment. Electronmicrographs such as those of Plate 3 do not show the situation in resting muscle, for in the course of preparing the sections the ATP diffuses out, causing the muscle to go into a state of *rigor* (see p. 13). The resulting strong bonds between actin and myosin can actually be seen in the electronmicrographs.

The only method available for examining the ultrastructure of muscle *while it is alive* is by means of x-ray diffraction. The principle is exactly similar to that of light diffraction, explained on p. 19. A narrow beam of x-rays of uniform wavelength, is directed at the living muscle. Most of this beam passes straight through the muscle and is absorbed by a small lead stop on the far side. Some of the rays are, however, diffracted by regions of high electron density within the muscle and fall on a photographic plate, where they form a characteristic pattern that is related to the periodicities in the ultrastructure of the muscle. The x-rays used typically have wavelengths of 1·5 Å rather than the 6000 Å of visible light. Consideration of Fig. 3.3 then shows that the spacings detected can be very small, right down to inter-atomic distances within a crystal. Of greater interest for our present purpose are the spacings of a few tens or hundreds of Ångstroms which arise from the ultrastructure of the filaments. These long spacings give only small angular deflections and demand corresponding experimental skill if the weak diffracted beam is to be distinguished from the far stronger undeviated one. The first successful small angle diffraction studies were made by H. E. Huxley around 1950: the chief, and rather surprising, result was that the longitudinal pattern (unlike that found by diffraction of visible light) did not alter when the muscle was stretched. This was the first piece of evidence to suggest that the protein chains must be capable of sliding past each other without being much distorted—one of the foundations of the sliding filament theory.

Recent advances in x-ray technique have led to diffraction pictures containing a very great deal of detail, as shown in Plate 2; so much detail, in fact, that not all of it has yet been interpreted. However, the following features of the pictures appear to be well established:

(i) The transverse diffractions on the equator (Plate 2 (a)) arise from the regular hexagonal arrangement of the thick filaments.

(ii) The diffractions along the longitudinal axis arise from the periodicities along the myosin and actin filaments, as indicated in Plate 2(b).

(iii) To either side of the equator, and more or less parallel with it, run *layer lines* which arise from the regular helical arrangement of cross-bridges on the myosin filaments. Accordingly, it can be seen that in resting muscle the layer lines have the same 435 Å periodicity as the myosin filaments. The lines visible at 217, 145 and 109 Å are higher orders of this spacing.

3·4   Changes in x-ray diffraction associated with functional change in the muscle

3.4.1   Change in length
As mentioned above, this does not lead to any change in the longitudinal diffraction pattern, indicating that actin and myosin filaments slide over one another without deformation. However, the equatorial spacings do change because when the muscle is stretched, it does not change in volume, so the filaments become more closely packed together.

3.4.2   Rigor
When the muscle goes into rigor the layer lines—derived from the cross-bridges—change their spacing to about 370 Å, close to that of the actin filaments. Probably this means that the cross-bridges actually become attached to the actin filaments.

3.4.3   Contraction
Recent advances in x-ray technique have reduced the total exposure time so that only ten minutes—instead of several hours—are now needed to expose a picture. A skeletal muscle cannot be held in a steady state of contraction for this period of time because of fatigue; but it has proved possible to build up the required total exposure by taking about one second from each of 600 short contractions. It is then found that during contraction the layer lines disappear altogether, presumably because the cross-bridges are in motion and thus no longer arranged in a regular pattern. The equatorial and axial reflections remain in their usual places.

Muscular contraction

In this chapter we shall move away from the minute details of molecular structure and consider what happens to whole muscles—or, at least, to whole muscle fibres—when they contract. The primary function of muscle is mechanical: it must be able to develop a controllable force and also to perform mechanical work by shortening against a force. A secondary function of muscles arises from the fact that they can generate a great deal of heat which can be vital for maintaining the temperature of warm-blooded animals or raising that of cold-blooded ones.

4.1 Experiments on living muscle

The first part of this chapter deals with experimental methods. If you mainly want to know about results, rather than about the way in which they are obtained, jump to p. 28.

4.1.1 Survival of tissue

Muscle is a hardy tissue and it will go on living for quite a long time outside the body if a few simple precautions are taken. It must be prevented from drying by being kept in contact with a 'physiological solution' of suitable osmotic strength and ionic composition. Most muscles contain an ample store of food substances, but some others, e.g. the smooth taenia coli of the intestine, must be given a continuous supply of glucose. The 'physiological solution' takes the place of the tissue fluid in which the muscle was bathed inside the animal's body, and it should be made up with roughly the same concentrations of inorganic salts as are found in the animal's blood. The solution need not be very complicated. For example, in a solution containing only NaCl 115 mM, KCl 2·5 mM and $CaCl_2$ 2 mM, a frog's muscle will live for two weeks if bacterial attack can be combated.

To provide an adequate oxygen supply is far more difficult. Only if the muscle is very thin, certainly less than 1 mm thick, can one depend on simple diffusion of oxygen from the outside of the muscle (and then only if the solution is kept saturated with oxygen rather than with air). If the central core of the muscle becomes severely short of oxygen during activity, it will soon cease to contract and may die. In order to experiment on large muscles such as those of mammals, it is thus essential to preserve an intact circulation of oxygenated blood.

4.1.2 Stimulation

For experimental purposes the muscle is usually stimulated by applying a brief electric shock, though for some purposes, and especially with smooth muscles, chemical stimulation may be employed instead. For example, potassium chloride, acetyl choline and caffeine can all produce contraction, and they do so by acting on entirely different parts of the excitable mechanism as will be explained in the next chapter.

Nevertheless, electrical stimulation is by far the most convenient to use for most purposes, as the brief electric pulses needed, of duration from 0·1 to 10 msec, can be so easily generated and controlled. Three cheap and simple circuits are shown in Fig. 4–1: if repeated pulses are needed the changeover switch must be replaced by a relay driven by an oscillator. If

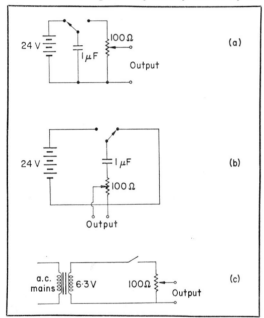

Fig. 4–1   Simple form of stimulator: (a) gives one shock each time the key is operated; (b) gives two shocks, in opposite directions, thus helping to minimize electrolysis and polarization; (c) gives continuous stimulation.

single shocks are not needed, but only repetitive stimulation, 50 c/s current derived from the mains (Fig. 4–1 (c)) is simple to use. In order to perform more elaborate experiments an electronic stimulator is required, operated either by valves or by transistors. According to its degree of complexity this may not only vary the strength and duration of the output pulses, but also deliver them in a preset pattern.

4.1.3   Stimulating electrodes

The electric shocks are led to the nerve supplying the muscle (indirect stimulation) or to the muscle itself (direct stimulation) via stimulating electrodes. Even if the shock is applied directly to the surface of the muscle it will probably operate by activating the nerve twigs within the muscle rather than by stimulating the muscle fibres directly unless the precaution has been taken of blocking neuromuscular transmission with curare.

The stimulating electrodes present an interface between the metal of the electric circuit and the salty solution bathing the muscle; electrolysis of the solution may occur at this interface with the formation of products that are damaging to the muscle. For this reason, copper or brass electrodes should be avoided since Cu^{2+} ions are toxic. Silver wires coated with silver chloride are satisfactory, since all the current is carried by Cl ions, but only if the current to be passed is very small. If possible, it is best to use inert electrodes of platinum: graphite such as is used for the electrodes of dry batteries is also satisfactory and very cheap.

4.2 Apparatus for recording mechanical changes

Two types of mechanical recording are commonly made. In one, *isometric* recording, the length of the muscle is held as constant as possible, and the variation in tension is recorded. In the other, *isotonic* recording, the tension is held constant and the muscle's alterations in length are recorded. There is no special magic about these two types of recording— they merely represent a convenient, though arbitrary, way of examining two aspects of the muscle's behaviour, and the records produced are closely related to one another (Fig. 4–8, p. 36). For the past century these recordings have been made by attaching the muscle to a suitable lever whose tip wrote directly on smoked paper. Simple apparatus like this is still very useful in teaching and, intelligently used, for some research purposes. However, for the highest performance it is essential to use *transducers* which transform the mechanical variable, length or tension change, into a proportional electrical signal, which can be recorded using an oscilloscope or an electrically driven pen recorder.

A typical purely mechanical isometric lever is shown in Fig. 4–2 (a). The writing lever is attached rigidly to a flat piece of spring steel so that when the muscle exerts a force, the torsion band is twisted and the writing point deflected. In an isotonic lever (Fig. 4–2 (b)) the lever is pivoted freely, if possible in ball or jewelled bearings; and facilities are provided for applying various loads close to the axis (so as to minimize their effective inertia). The initial length of the muscle can be controlled by an adjustable after-load stop. This type of lever may be equipped with a transducer instead of a writing point, for example, by causing the lever to interrupt part of a beam of light falling on a photocell.

The isometric lever may be modified similarly, but it is more satisfactory to employ an entirely different type of transducer such as is shown in Fig. 4–2 (c). Probably the most satisfactory tension transducer available nowadays is the silicon semiconductor strain gauge, whose electrical resistance is changed by minute changes in the length of the gauge. Two

of these, bonded to either side of a piece of steel of suitable size and
connected up as a Wheatstone's bridge, make an admirable tension
transducer. Best of all, if the gauges are bonded to the top and bottom
edges of the isotonic lever as shown by the dark strips in Fig. 4–2 (b) (an

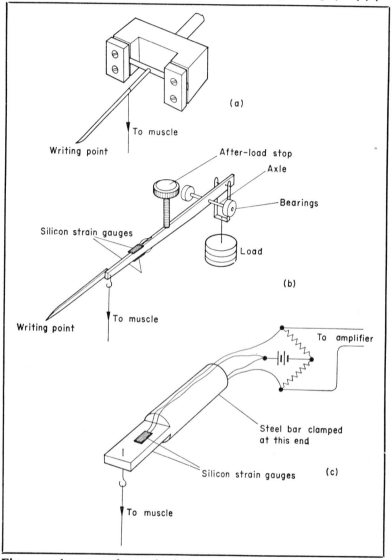

Fig. 4–2 Apparatus for mechanical recording: **(a)** Mechanical isometric
lever. **(b)** Combined isotonic and isometric lever. **(c)** Isometric transducer
employing silicon strain gauges in Wheatstone's bridge circuit.

arrangement first suggested by Dr. R. C. Woledge) both tension and length can be recorded simultaneously by a single lever.

4.2.1   Characteristics of performance

In assessing the performance of these or of *any* other recording instruments, three considerations are important—sensitivity, stability and frequency response. The frequency response is concerned with the accuracy with which the recorder can follow rapid changes. Many recording devices, such as the isometric levers shown in Fig. 4–2 (a) and (c) have a natural frequency of their own and this sets an upper limit to the performance. The limiting frequency is much higher for (c) than for (a).

Another important factor is the degree of damping, or friction, as illustrated in Fig. 4–3. When a force is suddenly applied to the lever, or removed from it, as shown in the upper line, the movement of the lever

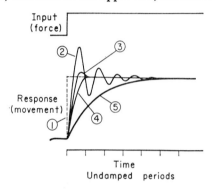

Fig. 4–3 The response of instruments such as the isometric recorders of Fig. 4–2 to a sudden step input. The interrupted line (1) shows how a perfectly undistorted response would appear. Curves (2)–(5) show the effect of progressively more and more damping. (From STARLING and LOVATT EVANS, 1962, *Principles of Human Physiology* (13th ed.). Eds. DAVSON and EGGLETON. Churchill, London. Fig. V.23 (A)).

does not immediately follow the change in force; thus it does not follow line 1, but one of the others. If the damping is too small, the lever will oscillate (line 2); if damping is too great the lever will creep slowly to its final position (line 5). In either case much time is lost before the lever stops moving. The damping should therefore be adjusted to some intermediate value (4 or preferably 3); the lever then settles down in a time roughly equal to one (undamped) period. It can be shown that the lever then responds most accurately not only to the step input that we have considered so far, but also to most other patterns of input too. A short natural period of oscillation is clearly a desirable characteristic; unfortunately it can only be obtained at the price of a stiff spring and correspondingly low sensitivity. As in most things, a compromise is necessary.

An isotonic lever has no natural period, only an inertia, whose effect at the point where the muscle is attached should be kept as small as possible.

4.3 The relation between stimulus and response

A typical striated muscle responds to a single adequate stimulus by giving a *twitch*, that is, a brief period of contraction followed by relaxation. As indicated in Fig. 4–4 (a), the time course of the twitch depends on the particular type of muscle examined, and in a single type of muscle, it depends on the temperature. As with many other biological—and chemical

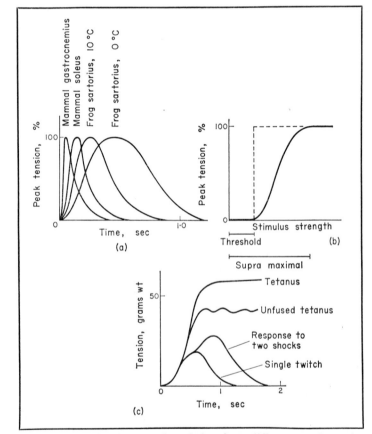

Fig. 4–4 Graph showing muscle twitches from different species and temperatures, scaled to the same peak height. **(b)** The relation between strength of stimulus and size of response. The interrupted line shows an all-or-nothing response, the continuous line a graded one. **(c)** The summation of response following repeated stimulation. Frog muscle 0°C.

—processes, a 10°C rise of temperature increases the speed between two and three times.

The size of the twitch response depends on the strength of the stimulus, as shown in Fig. 4–4 (b). With very weak stimuli, nothing happens; but when the strength passes the *threshold* a small response is found which increases progressively until the stimulus reaches its *maximal* value. This effect arises because the weak shock stimulates only a few muscle fibres close to the electrodes where the current density is highest, while the supramaximal shock stimulates all of them. Clearly, in experiments on the mechanical properties of muscle (and most other properties too), care must always be taken to employ only supramaximal shocks, otherwise it will be impossible to obtain consistent results.

It will be seen in the next chapter that the response of each individual muscle fibre is not graded, but is of the *all-or-nothing* type indicated by the interrupted line: if a shock is strong enough to produce a response at all, that response will be maximal. Tissues that show this type of behaviour are sometimes said to obey the all-or-nothing *law*, but there is really very little law about it. In many types of muscle even the individual fibres show graded responses. On the other hand, in heart muscle the whole piece of tissue shows all-or-nothing characteristics, because excitation can spread freely from one muscle cell to another. It is important to be clear about one thing that the all-or-nothing 'law' does *not* say. It does not claim that all the responses must be of the same size; they may diminish in size as a result of fatigue, or they may actually increase in size as a result of previous stimulation, an effect called facilitation. The critical point is that the size of the response, no matter what it happens to be at the moment in question, cannot be increased by increasing the strength of the stimulus.

4.3.1 Repetitive stimulation: twitch and tetanus

If a second shock is given to the muscle before the response to the first has completely died away, *summation* occurs, as shown in Fig. 4–4 (c). If the stimuli are repeated regularly at a high enough frequency, the result is a smooth *tetanus*, with tension maintained at a high level, as long as the train of stimuli continues, or until fatigue comes on.

4.4 The tension–length curve

Resting muscle is elastic. It can only be stretched by applying a force, and the greater the force, the greater the extension, as shown by the lower curves in Fig. 4–5 (a), (b) and (c). These also show that the muscle does not obey Hooke's law in that it becomes more and more inextensible the further it is stretched, i.e. the curves become steeper and steeper. The resting elasticity results largely from the meshwork of connective tissue within the muscle, whose fibres become progressively taut when the muscle is stretched. Exactly the same effect is seen when a knitted stocking is stretched. The connective tissue is mechanically in parallel with the contractile component (CC), as indicated in Fig. 4–5 (d), so it constitutes a parallel elastic component (PEC): there are also elastic structures, partly in the tendons, that constitute a series elastic component (SEC). When muscle is stimulated, the CC develops a tension which varies according to the length of the muscle in the way shown by the d curves in Fig. 4–5 (a), (b) and (c). Since CC and PEC are in parallel, their tensions must be added together, so what is actually recorded in each case is shown by the

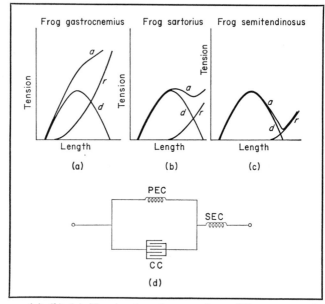

Fig. 4–5   (a), (b), (c) Tension-length curves from different types of muscle. In each case r shows the tension–length curve of the resting muscle, a shows that of the active muscle and d $(=a-r)$ shows the extra tension developed on stimulation. Note that even if d and r are the same in all muscles, the shape of a varies widely, depending on the length at which resting tension begins to be developed. (d) Equivalent mechanical components in muscle, CC= contractile component; PEC=parallel elastic component; SEC=series elastic component.

upper full curve a. The great difference in the shape of the a curve between Fig. 4–5 (a) and (c)—notably the absence of a dip in (a)—merely arises from variation in the overlap of the CC and PEC curves, and thus from the amount and distribution of the internal connective tissue.

Note that all these curves are obtained by setting the length of the muscle *before* it is stimulated. If the length is changed *while* the muscle is being stimulated, a different and more complex relation between tension and length is found.

The maximum tension developed on tetanic stimulation varies from 1·5 to 4·0 kg/cm² (frog, mammal) up to about 10 kg/cm² (edible mussel). When describing the tension development of a particular muscle, it is essential to express the result 'per square centimetre of cross-sectional area', otherwise comparison between different muscles is meaningless. The length (L) cm and weight (M) grams of a muscle are more easily measured than its cross-sectional area, so if the cross-section is fairly uniform it is convenient to estimate its area as M/L, thus assuming a density of 1 g/cm³, which is approximately correct.

4.5 The tension–length curve and the sliding filament theory

One of the earliest predictions from the sliding filament theory of contraction was that when a muscle was lengthened, the area of overlap of thick and thin filaments should diminish, so the tension developed should diminish also. It had been known for about a century that this did indeed happen (see the right-hand ends of the d curves in Fig. 4–5); but to make an accurate quantitative comparison between tension development and area of overlap has proved far more difficult. Such a comparison demands: (1) accurate measurements of the lengths of the thick and thin filaments and of their overlap at various sarcomere lengths, and (2) measurements of the tension–length curve which were accurately related to the sarcomere length rather than to the length of the muscle as a whole.

The first requirement has been met by taking electronmicrographs of muscles that had been fixed and sectioned with especial care to avoid artifacts from shrinkage and other cases. The results are summarized in Fig. 4–6 (a).

The second requirement has proved far more difficult to satisfy. Using a whole muscle, even an extensible one like the semitendinosus, the tension–length curve at long lengths depends to an inconvenient degree on the connective tissue present. When this problem was eliminated by working with a single fibre freed of connective tissue, another problem became apparent. The sarcomeres at the ends of the fibre were shorter than those in the middle, so it was impossible to ascribe the tension developed by the fibre to one particular sarcomere length. The final solution, employed in obtaining Fig. 4–6 (b), was to make the measurements on the middle part only of a single fibre, effectively eliminating the unwanted contributions of the ends of the fibre by a sophisticated electromechanical feedback system. It then appears that the tension–length curve consists of straight segments with fairly sharp corners between them. In the whole fibre, and even more, in the whole muscle, these sharp corners become rounded because of the non-uniformities mentioned above. Moreover, as illustrated in Fig. 4–6 (c), the positions of the corners correspond in an intelligible fashion with the various stages of overlap of the filaments. The fall in tension at the left-hand side of the curve is not so

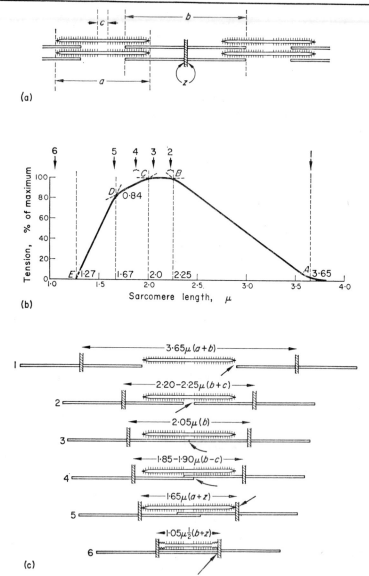

Fig. 4–6   (a) Standard filament lengths. $a = 1\cdot60\ \mu$; $b = 2\cdot05\ \mu$; $c = 0\cdot15\text{–}2\ \mu$; $z = 0\cdot05\ \mu$. (b) Tension–length curve from part of a single muscle fibre (schematic summary of results). The arrows along the top show the various critical stages of overlap that are portrayed in (c). (c) Critical stages in the increase of overlap between thick and thin filaments as a sarcomere shortens. (From A. M. GORDON, A. F. HUXLEY and F. J. JULIAN, 1966, *J. Physiol.*, **184**, 170–192.)

easy to explain as that on the right. It appears that extensive overlap of the filaments interferes with the formation of cross-bridges, while the rigidity of the thick filaments themselves probably absorbs part of the force that has been developed.

4.6 Tension–length curves in other types of muscle

Tension–length curves of roughly the same type, with tension development rising to a maximum and then falling again, are found in almost all types of muscle, cardiac and smooth as well as skeletal. The outstanding

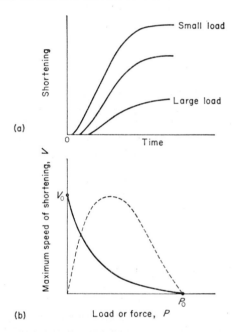

Fig. 4-7 **(a)** Records of shortening against time from a muscle lifting various loads. Tetanic stimulation started at time zero. **(b)** Force–velocity curve. The interrupted line shows how mechanical power produced (=force × velocity) varies with the force on the muscle. (After STARLING and LOVATT EVANS, 1962, *Principles of Human Physiology* (13th ed.). Eds. DAVSON and EGGLETON. Churchill, London. Fig. V.29 (A).)

exceptions are the insect fibrillar muscles in which the thick filaments extend the full length of the sarcomeres. These muscles are damaged if they are extended by more than a few per cent and they normally work not so much by developing a steady tension as by exhibiting a negative resistance, i.e. while they are being stretched the force that they exert diminishes. This enables them to vibrate continuously if they are coupled up to a suitable mechanical system containing inertia and elasticity such as is normally provided by the insect's thorax and wings.

It is only in vertebrate skeletal muscles that the ultrastructural basis of the tension–length curve has been demonstrated. Presumably a similar mechanism is operating in other types of muscle though at present it is hard to visualize what is happening in those vertebrate smooth visceral muscles in which no system of filaments has yet been demonstrated: perhaps the actin and myosin present form into filaments only when the muscle is actually contracting. This might account for the *plasticity* that many smooth muscles possess—the property that enables them to develop maximum tension at various lengths according to their previous treatment, and thus to shift the whole tension–length curve back and forth along the horizontal axis.

4.7   Isotonic contraction

If a muscle is attached to a suitable isotonic lever (see Fig. 4–2 (b), tetanized, and allowed to lift various different loads, a set of records similar to Fig. 4–7 (a) is obtained. It can be seen straight away that the greater the load, the smaller the total shortening. In fact, force and final length follow the left-hand part of the tension–length curve for active muscle (*a*) that was described in the previous section and shown in Fig. 4–5. Another striking effect is that the greater the load, the smaller is the maximum *speed* of shortening, i.e. the less is the maximum slope of the curves. If force is plotted against slope a graph similar to Fig. 4–7 (b) is obtained.

4.7.1   The force–velocity curve

This relation between force and velocity is an important characteristic of contracting muscles. Curves similar to Fig. 4–7 (b) have been obtained from all the muscles that have so far been examined—cardiac and smooth as well as striated—and even from contracting actomyosin threads. The only exceptions seem to be those insect muscles that work by small-scale vibrations rather than by gross changes in length. Much thought has therefore been expended in the effort to find out what it is that makes muscles behave in this way. One early idea was that the force developed by the muscle was actually constant at all speeds, but that when the muscle shortened, part of the force was absorbed by an internal viscosity, so that less force was left over to be exerted externally. If the internal viscosity was a non-linear one, this theory would account perfectly well for the purely mechanical properties of muscle; but it was abandoned because it did not seem to fit with the observed energetic properties of contraction. It now seems probable on various grounds that the curve is an expression of the fact that the rate of the chemical reactions in the muscle is somehow linked to the force on it.

Many different equations can be fitted accurately to the force–velocity curve. The most interesting one is Hill's equation (HILL, 1938) which fits part of a hyperbola to the curve.

$$V = (P_0 - P)b/(P + a)$$

where V is velocity, P is force acting, P_0 is the isometric force and a and b are constants. This can also be written in terms of the relative speed

$$V' = (1 - P')/(1 + P'/k)$$

where $V' = V/V_0$, $P' = P/P_0$, $k = a/P_0 = b/V_0$. V_0 is the maximum (unloaded) speed.

One practical consequence of the shape of the force–velocity curve is that the mechanical power output of the muscle ($= P \times V$, interrupted line in Fig. 4–7 (b)) passes through a maximum when the force on the muscle and its speed have about one third of their maximal values. For efficient performance of mechanical work, therefore, it is necessary to arrange that the load presented to the muscles has about this value. A three-speed gear on a bicycle is an excellent example of a practical device that makes it possible to match load and speed to the properties of the muscle regardless of the incline. Incidentally, the maximal power output is about $0.1 P_0 \times V_0$.

4.7.2   Mechanical components in muscle

It will be recalled from Fig. 4–5 that muscle contains an elastic element in series with the contractile one. This series elastic component (SEC) is partly in the tendons and partly distributed along the sarcomeres. It has a very important effect in modifying the pattern of contraction. Even if the muscle as a whole is held rigidly isometric, when it becomes active there will still be some internal shortening of the contractile element with corresponding stretching out of the SEC. This internal shortening slows down the rate at which tension rises in the muscle and this is the reason why tension does not rise so high in a twitch as in a tetanus even though the change in the contractile proteins seems to be the same in both; activity in the twitch begins to decay before the tension has had time to reach its full value. Since the peak tension attained in a twitch depends on a balance between two processes, it is no surprise to find that the twitch : tetanus ratio varies greatly from one type of muscle to another, and can be greatly diminished by introducing extra compliance into the recording system.

4.7.3   Isometric versus isotonic

This is perhaps a suitable point at which to re-emphasize that these are simply two out of several possible methods of *recording*; the change in the contractile component need not be very different in the two cases. In every type of contraction it is important to realize that a sequence of changes in *both* length and tension is taking place. Figure 4–8 indicates these changes as they occur in after-loaded isotonic twitches (i.e. starting with the load supported on a stop). It is clear that as the load is increased, the muscle simply spends more and more of its time contracting isometrically.

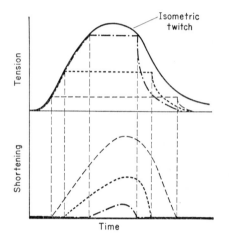

Fig. 4-8 Changes in length and tension recorded simultaneously (using a lever such as that shown in Fig. 4–2**b**) during after-loaded isotonic twitches against various loads.

4.8 The isolated contractile system

So far we have dealt with the properties of whole living muscles excited in the normal way through their cell membranes. In the last twenty years a great deal has been learnt about the molecular basis of contraction by experimenting with simplified contractile systems that contain some, but not all, of the constituents of living muscle. The most useful preparation of this type, originated by Albert Szent-Györgyi, is made by soaking strips of freshly dissected muscle in a buffered aqueous solution of glycerol. This apparently destroys the cell membranes and allows the diffusible contents, including ATP and PC, to diffuse away; but it leaves the contractile proteins *in situ* in their normal highly ordered state. After the treatment with glycerol, the fibres are in a state of rigor, because all the ATP has diffused away (see p. 13). When ATP is added in the presence of calcium and magnesium ions, it is actively hydrolysed to ADP by the actomyosin ATP-ase present. At the same time the fibres contract, developing tension, shortening and performing mechanical work. The activated fibres exhibit a normal tension–length curve and a force–velocity curve of normal shape. The actual speed of shortening is less than in living muscle, but this seems to be caused largely by the rather slow diffusion of ATP into the interior of the fibres.

The active fibres can be caused to relax again by removing Ca^{2+} from the system, since they normally contain the extra proteins that sensitize actomyosin to Ca. The Ca is most simply removed by adding etheylene glycol bis (β-amino ethyl ether)—N,N'—tetra acetate (EGTA), a compound which binds calcium very strongly. When the concentration of ionized calcium falls below about 10^{-7} M, the ATP can no longer be hydrolysed, and active contraction stops. However, the fibre does not revert to its original state of rigor, because ATP is present and this keeps actin and myosin from becoming stuck together (see p. 13). The fibre remains extensible, just like a living fibre at rest. There is now evidence that this calcium-operated control mechanism demands the presence of other proteins besides actin and myosin; the third protein resembles tropomyosin B. The control mechanism of the intact muscle probably works in a similar way, by active control of the concentration of ionized calcium within the muscle cell.

Glycerol-extracted preparations have been made from a wide variety of striated and smooth muscles and they all seem to function in essentially the same way. The exceptions, as usual, are the insect fibrillar muscles. When ATP is added to them they can, if suitably loaded, generate long-lasting vibrations.

4.9 Theories of contraction

There no longer seems to be much room for doubt about the essential features of the sliding filament theory, but we still have very little idea how the cross-bridges actually produce force and the continuous motion of one type of filament over the over. At present there are two main theories about the way in which the cross-bridges might operate.

In the first theory (A. F. HUXLEY, 1957) the cross-bridges attached to the myosin are thought of as flexible, with active groups at their ends: such structures would be small enough to be continuously agitated by Brownian movement. It is proposed that the active groups can become attached to the actin and that their probability of doing so is higher in the 'upstream' than in the 'downstream' direction. Elaboration of this idea, and its mathematical formulation, makes it possible to predict quantitatively several of the mechanical and energetic properties of muscle such as the force–velocity curve and the pattern of heat production under various conditions. However, there is at present no way of testing the theory at ultramicroscopic level.

In the second theory (R. E. DAVIES, 1963) the cross-bridges are thought to perform in a more organized fashion, rather than depending on Brownian motion. At rest the bridges are supposed to be held in an extended form by electrostatic repulsion between an ionized ATP^- bound to the tip and a fixed negative charge at the base, where the bridge is attached to the myosin filament. The Ca^{2+} ions released during activation are supposed to form links by electrostatic attraction between the charged tips of the bridges and negatively charged sites on the actin filaments. Actin and myosin thus become attached through a calcium link; at the same time the charges whose repulsion was holding the bridge extended have been neutralized, so the bridge shortens and pulls the actin filament along the myosin filament. This cycle is repeated several times over during the course of a single contraction, thus producing continuous motion. Davies' theory does not lend itself so easily as A. F. Huxley's to mathematical formulation and quantitative testing. On the other hand, it does link together a number of facts about the chemistry of contraction and suggest fresh lines of investigation. What is needed is a new theory that combines the virtues of both old ones; and more important still, more basic knowledge about the physical chemistry of such very small but highly ordered structures as the cross-bridges.

The control system 5

Muscles are not much use to an animal unless their contraction can be controlled in accordance with the needs of the organism as a whole. The type of control mechanism required naturally depends on what particular job the muscle is supposed to be doing. For example, the heart muscle must keep on beating throughout life and its muscle fibres have an intrinsic contractile rhythm—whose mechanism will be described later—so that the heart goes on pumping even if it is completely separated from the rest of the body. The nerves that connect the heart with the brain serve only to modify the beat, they do not initiate it. At the other end of the scale are the skeletal muscles which *never* contract unless they receive the order to do so from the central nervous system, and whose pattern of response must follow very accurately, and very swiftly, the set of instructions sent out from brain or spinal cord.

Contraction is fundamentally a biochemical process, and there is now good evidence that it is controlled biochemically within the muscle cell by active regulation of the concentration of ionized calcium. However, chemical control of this kind suffers from one fundamental limitation. Change of concentration involves diffusion, and the time required for diffusion increases roughly as the square of the distance over which it must operate. In practical terms, it is only possible to reduce switching time to a few milliseconds if the distance for diffusion does not exceed a few μ. Thus for a striated muscle fibre, diffusion will suffice to control contraction within the length of a single sarcomere (say 2 μ) but not to transmit the order to contract from the surface of the fibre to its centre (say 50 μ)—let alone from the spinal cord to the muscle, which may be a distance of a metre.

For quick transmission of information over long distances an electrical system is therefore employed. Unlike comparable man-made systems, this one must be made out of materials that are not particularly good conductors of electricity, for the currents are carried by ions rather than by electrons as in a metal. To make matters worse, there are no very good insulators either. For these and other reasons it would not be practicable to use a simple electric circuit to pass currents from the brain down the nerves to the muscles.

5.1 The role of the cell surface

The electrical potentials used in signalling are generated at the surface of the nerve and muscle cells because the cell membranes are not uniformly permeable to all the ions present and these ions are not uniformly distributed on both sides of the membrane. This leads, by a mechanism

39

that will be described later, to a resting potential difference of about 50 to 100 mV, the inside of the cell being negative to the outside. In all types of muscle the contractile state is linked to the potential across the membrane and if the potential difference is reduced, the muscle fibre tends to go into contraction. Many other types of cell also show resting potentials, but only in nerve and muscle cells, and some sensory cells, is their functional importance known.

5.1.1   Electrical recording methods: microelectrodes

The only really satisfactory way of recording membrane potentials is by placing one electrode actually inside the cell. Suitable electrodes are made by drawing glass tubes out into *very* fine tips less than 0·5 μ in diameter. Such a tip is so fine that it cannot be seen in an ordinary microscope because it is smaller than the wavelength of light, but because of its fineness it is able to penetrate the cell membrane without damaging it. The arrangement of recording apparatus is shown in Fig. 5–1. A pathway for electrical

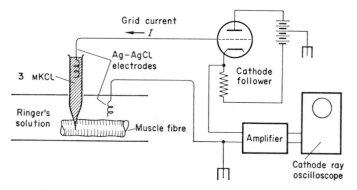

Fig. 5–1 Arrangement of apparatus for internal recording, using micro-electrodes. (From STARLING and LOVATT EVANS, 1962, *Principles of Human Physiology* (13th ed.). Eds. DAVSON and EGGLETON. Churchill, London. Fig. V.10.)

conduction through the microelectrode is provided by filling the capillary completely with an electrolytic solution such as 3 M KCl. This solution has a high conductance, but there is so little of it present in the electrode tip that the total resistance of the electrode is quite high, usually 5 to 20 megohms. This high resistance poses a special problem in electrical recording as even very small spurious currents passing through the electrode, such as the grid current I, will give rise to relatively large spurious potentials. Hence the use of the cathode-follower input stage shown in Fig. 5–1: this circuit does not amplify the signal, indeed it actually diminishes it somewhat, but it does have a very high input resistance.

5.2 The origin of membrane potentials: the Nernst equation

The concentration of ions inside and outside a muscle fibre are shown somewhat diagrammatically in Fig. 5–2 (a). In order to see how this distribution of ions gives rise to a potential difference, first imagine that the cell membrane is selectively permeable to K^+ ions only. At first some

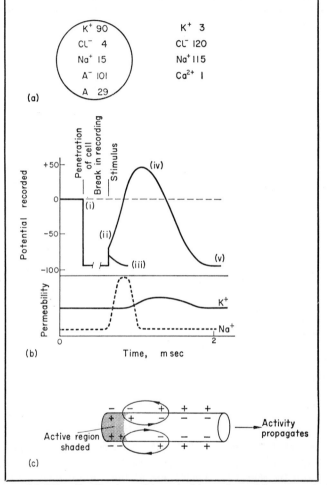

Fig. 5–2 (a) The distribution of ions inside and outside a muscle fibre. Approximate concentrations in gram-ions per litre. A' and A are non-diffusible organic substances, ionized and non-ionized respectively. (b) The changes in potential recorded by a microelectrode inside a muscle fibre (upper curves). The lower curves show the underlying changes in membrane permeability. (c) The currents generated by an active region flow in such a way as to stimulate neighbouring resting regions.

of them will tend to diffuse down the concentration gradient and out of the cell. However, when they do so, they will leave behind a net negative charge. Thus each K^+ ion is subject to two opposing forces. The loss of only a few positive ions, far fewer than can be detected by chemical analysis, will suffice to charge up the electrical capacitance of the cell, and the resulting difference of potential tends to impede further loss of ions. The system will come to equilibrium when the two opposing tendencies are just equal, that is, when a potassium ion neither gains nor loses energy if it leaves the cell. For purposes of calculation, imagine that one mole of K^+ ions leave the cell. Then, energy gained from the fall in concentration, by analogy with that obtained from expansion of a perfect gas,

$$= RT \log_e \text{(conc. inside/conc. outside)}$$

while energy given up in overcoming the potential gradient

$$= E.F$$

where E is the membrane potential and F (Faraday's constant) is the total charge carried by 1 mole of a univalent ion. At equilibrium the energy gained is just equal to that given up, so

$$E = \frac{RT}{F} \cdot \log_e \frac{\text{conc. outside}}{\text{conc. inside}}$$

This is a very important equation, often called the Nernst equation, which can be put into a more useful form by inserting the values of the constants for $T = 25°C$, and converting to $\log_{10}$; then

$$E = 59 \cdot 1 \log_{10} \text{(conc. inside/conc. outside) millivolts}$$

Thus in the example shown in Fig. 5–2 (a) where the concentration ratio is 30:1, the resting potential would be 87·4 mV.

Now in fact the cell membrane is not simply permeable to K^+ ions but also to chloride ions and several others. Clearly, exactly the same equation must apply to all these permeant ions. Moreover, the membrane potential can only have one value, which must be the same for all permeant ions, so at equilibrium

$$(K_{out}/K_{in}) = (Cl_{in}/Cl_{out}), \text{ etc.}$$

It will be noted in Fig. 5–2 (a) that the chloride ions do indeed have a concentration ratio of 30:1. However, it is plain that the sodium ions are *not* following the same rule. Their concentration ratio is such as to set up a membrane potential of 52·3 mV, in the opposite direction, with the inside of the cell *positive*. The reason why this potential does not appear is that the membrane is very impermeable to sodium ions and those that do enter are pumped out again by an active process—the sodium pump. It will be noted that there are also organic negative ions (A′) within the cell which cannot pass through the membrane. The presence of impermeant ions, and the need for both electrical and osmotic equilibrium sets up, by purely physico-chemical means, what is called a Donnan equilibrium, and it is this that determines the resting concentration of ions shown in Fig. 5–2 (a).

The state of the contractile proteins is linked to the magnitude of the membrane potential: if the fibre is depolarized, i.e. if its inside becomes less negative, then the contractile proteins tend to go into the state of contraction. The mechanism responsible is only partly understood, and will be described later—at present we are more concerned with the means by which depolarization may be brought about.

5.2.1   Electrical

The cell membrane has a high electrical resistance, so when current passes through it, by Ohm's law a relatively large potential difference will be generated. If the current is passing from the inside of the cell to the outside, this potential difference will be in the opposite direction from the resting potential and will lead to depolarization and contraction. One common way of causing current to pass outwards through the membrane is to apply a negative pulse to a stimulating electrode on the surface of an isolated muscle. A similar effect is produced in the intact animal when an action potential passes along the surface, as we shall see later.

5.2.2   By altering the ion balance

If the concentration of potassium ion outside the fibre is increased, it follows from the Nernst equation that depolarization must follow. This is the mechanism by which potassium contracture is produced.

5.2.3   By altering the selective permeability of the membrane

Since the existence of the resting potential depends on selective permeability, anything that destroys this property will lead to depolarization. Mechanical damage certainly works in this way. More physiologically, the application of acetyl choline does the same under many circumstances.

Although the basic electrical mechanism seems to be similar in all different types of muscle, it is modified in detail so that it is adapted to their particular circumstances.

5.3   Skeletal muscle

In the striated muscle used by vertebrates for their quick movements, the whole fibre is rapidly depolarized by an action potential propagated over its surface. The way in which this process works is illustrated in Fig. 5–2 (b), which shows the type of record that can be obtained by intracellular recording from a single muscle fibre, using apparatus similar to that shown in Fig. 5–1. When the fibre is penetrated by the microelectrode (i), the potential recorded suddenly drops to approximately

—90 mV, its resting value. At (ii) the fibre has been suddenly depolarized by making a pulse of current flow outwards through the membrane. If the depolarization is only small, the potential soon returns on an exponential curve to its resting value (iii). However, if the applied depolarization exceeds a certain value, something entirely different happens; the membrane proceeds to depolarize itself and rapidly sets up a potential of 30–40 mV in the opposite direction (iv). Finally the potential returns, somewhat more slowly, to its resting value (v). This sequence of events is called an *action potential* and the bottom part of Fig. 5–2 (b) shows the underlying changes in the cell membrane that bring it about. At rest, as we have already seen, the permeability to K^+ is high, while to Na^+ it is low. When an adequate stimulus is given, the Na^+ permeability suddenly rises—as a result of changes in the membrane that are still not understood—to such an extent that a sodium potential is set up and the inside of the fibre becomes temporarily positive in relation to the outside. The rise in K^+ permeability that follows speeds up repolarization.

So far we have been considering what happens at a single point on the fibre. Figure 5–2 (c) shows how an action potential set up at one point on a fibre will lead to a circulation of electric current inside and outside the fibre, the result of which is to drive current outwards through the membrane in neighbouring regions. These in turn become depolarized and generate their own local action potentials, so the disturbance is propagated along the fibre. Essentially the same process happens in nerve fibres as in propagating muscle fibres.

The order to contract is transmitted from the central nervous system by an action potential passing down the fine branch of a motor nerve. The amount of current generated by the fine nerve branch is too small to stimulate the much larger muscle fibre directly but at the neuromuscular junction is a specialized *end-plate* which acts as a kind of amplifier. When the nerve impulse reaches the end-plate, some acetyl choline is released which effectively depolarizes the muscle membrane by increasing the permeability to all ions; this sets off an action potential which is propagated down the muscle fibre. In its wake follows a wave of contraction. It has been shown in recent years that the link between excitation and contraction is provided by a specialized internal conduction system within the muscle fibre. As shown in Fig. 5–3, the space between the myofibrils contains a ramifying system of tubules called the *endoplasmic reticulum*. Two parts of this system are almost certainly involved in excitation. The transverse tubules, sometimes called the T-system, start as openings on the surface of the fibres and run inwards along the Z-line. Every now and then they pass between paired *outer vesicles* forming characteristic-looking structures called *triads*. There is evidence to show that the outer vesicles have the power to pump up the calcium ions from the sarcoplasm, thus lowering the concentration below that at which ATP splitting can occur. In the resting muscle it is thought that they contain much of the free calcium. When an action potential sweeps over the surface of the fibre, some sort of electrical signal passes down the T-tubule and causes the Ca^{2+} to be released from the outer vesicles into the sarcoplasm, so ATP splitting and contraction are initiated. Unless another action potential comes

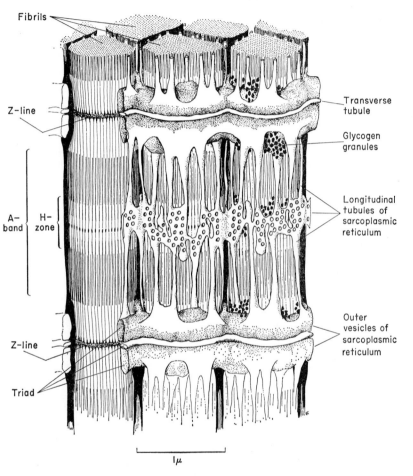

Fig. 5–3 The endoplasmic reticulum. (From PEACHEY, 1965. *Excerpta Medica International Congress Series* No. 87, p. 391.)

along, the Ca^{2+} is soon pumped back into the vesicles, and the muscle relaxes. In skeletal muscle the action potential lasts only about a millisecond, while the resulting muscle twitch lasts from ten to a thousand times as long. It is thus possible for the contractile system to be reactivated long before the contraction has begun to decline and this is

4+M.

exactly what happens when a muscle is tetanized. Caffeine, which produces contraction without altering the membrane potential, seems to act directly on the vesicles.

Not all skeletal muscle fibres are capable of conducting action potentials and therefore of responding by twitches and tetani. Some striated fibres which appear to be specialized to produce long-lasting tension are innervated by highly branched nerve fibres which make contact at many points of the fibre surface. The acetyl choline released produces a local and progressive depolarization with a corresponding graded contraction. Such fibres do not obey the all-or-nothing 'law' which, after all, merely reflects the all-or-nothing nature of the action potential.

5·4 Cardiac muscle

Although cardiac muscle is striated, the mechanism controlling contraction differs in at least four important respects from that of skeletal muscle:

1. The cells can conduct action potentials which arise by a mechanism similar to that previously described. However, as shown in Fig. 5–4, repolarization is very much delayed, so that the whole action potential lasts perhaps 100 msec instead of the 1 msec seen in skeletal muscle.

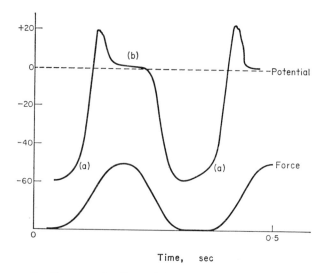

Fig. 5–4 Cardiac muscle. Potential changes recorded by an internal microelectrode and their relation to the tension developed (lower curve). Note the spontaneous depolarization (a) and the delayed repolarization (b): also the slow time scale when compared with Fig. 5–2.

2. As shown by the lower curve in Fig. 5–4, contraction lasts hardly any longer than the action potential, so it is impossible to tetanize cardiac muscle.

3. Cardiac muscle fibres, especially those from the atria and auricles, show a tendency to contract rhythmically without the need for external stimulation. This is because their resting potential is not stable: there is a tendency to depolarize spontaneously at a rate that is sufficiently high to trigger off the action potential.

4. Action potentials are freely conducted from one cell to another so that excitation of one part of the heart leads finally to contraction of all of it. In the absence of external stimulation it follows that the rhythmic excitation arising in the atria is conducted to the auricles and ventricles.

5.5 Smooth muscle

The smooth muscles form a very large group with widely varying properties, but in general smooth muscles are composed of small cells giving slow but sustained contractions; they also tend to be sensitive to such chemical agents as acetyl choline, adrenaline, histamine, 5-hydroxy-tryptamine and oxytocin.

Some types of smooth muscle resemble skeletal muscles in that they are connected by motor nerves to the central nervous system and normally respond chiefly to impulses coming down these nerves. However, since neuromuscular transmission is chemical, they may also be activated by local or general liberation of the appropriate agent. Examples of this type are the vasoconstrictor muscles in mammalian arterioles (adrenergic) and the muscles that retract the byssus—the fibrous attachment—of the edible mussel (cholinergic).

The muscles of the viscera more closely resemble cardiac muscle. Excitation can be conducted from one cell to another, as in the heart, but the excitability of the cells waxes and wanes so it does not necessarily follow that an impulse originated at one point will spread throughout the organ. Nerve fibres and even nerve cells are often present but they seem mainly to act (as in the heart) to modify activity rather than to originate it. The combination of spontaneous activity with conduction can lead to the generation of a regular series of waves such as those seen in the ureter.

Smooth muscle cells have a resting potential which is smaller and much more inconstant than that of skeletal muscle. The essential thing for contraction is depolarization, which may occur as a gradual shift of the mean membrane potential or as the result of the generation of action potentials. Action potentials seem to be a necessity for conduction to occur from one cell to another.

Probably calcium plays its usual role at biochemical level in controlling contraction but there are no specialized organelles like the triads to handle it. The muscle cells are so small, and their contraction so slow, that diffusion of Ca^{2+} from the cell surface would be quite quick enough for the purpose.

5.6   Invertebrate muscles

Many other types of neuromuscular apparatus are found and it is clear that the vertebrate arrangements that have just been described are not the only ones that can work effectively. For example, in some crustaceans a large muscle may be supplied by only two nerve axons, each of which branches to make contact with every muscle fibre. One axon is excitatory, the other inhibitory; and their combined use evidently leads to a quite sufficient degree of motor control.

In the locust, the powerful leg muscle is supplied by three axons. Impulses in one axon lead to propagated action potentials in the muscle fibres and the rapid twitches which the animal uses for jumping. Stimulation of the second axon gives slow, non-propagated contractions and the third axon serves to modify the responses induced by the other two.

The oscillating types of insect flight muscle are activated by propagated action potentials. Each action potential may lead to several oscillations since these arise from the peculiarity of the contractile system and not at all from the excitatory system as such.

Energy supply for contraction

The distinctive feature of the muscle proteins is their ability to transform energy from chemical to mechanical form. Ultimately the energy for muscular contraction, as for all the other processes of life, comes from chemical reaction between the foodstuffs that we eat and the oxygen that we breathe. However, we have seen that the only substance that is able to activate the contractile proteins is ATP. Accordingly, within the muscle a complex series of chemical reactions couple the oxidation of foodstuffs to the synthesis of ATP. There is no point at this stage in going into details about this chemical transformation, which is best thought of as a chain for transmitting chemical energy to the contractile proteins and for dividing up a large packet of energy into smaller and more manageable pieces.

6.1 Thermodynamic principles applied to muscle

The laws of thermodynamics impose clear restrictions on the possible transformations of energy. There are two distinct types of energy—heat and work. In the 'work' category we include mechanical work, electrical energy and chemical free energy. Under the first law of thermodynamics—the law of the conservation of energy—all these forms of energy are equivalent to each other so that the sum total of all of them must remain constant. It is when one actually tries to transform energy from one kind to another that the limited nature of this equivalence becomes plain. According to the second law of thermodynamics, in a system like a muscle *which is at uniform temperature*, the following restrictions apply:

1. It is very easy to convert work into heat. Friction, viscosity, electrical resistance—in all of them this conversion takes place.

2. It is possible to convert one form of work into another. For example, chemical free energy can be converted directly into electrical work in a galvanic cell, or into mechanical work in a muscle.

3. It is completely impossible to convert heat into work. This may seem somewhat surprising, as in a steam engine some heat is converted into work. This is only possible because the steam engine, unlike the muscle and almost all other biological systems, is *not at uniform temperature*: unless there is rather a large temperature gradient within the engine, only a very small fraction of the heat can be converted to work.

It follows from (1) and (3) that once energy has been converted to heat, there is no way of changing it back into work again, so this step is irreversible. Thus in the transformation of chemical energy into mechanical work (2), any free energy that is not successfully converted is degraded irreversibly to heat. The proportion of energy thus degraded measures the inefficiency of conversion: so

$$\text{efficiency} = \frac{\text{work obtained}}{\text{free energy made available}}$$

6.2   Chemical changes in muscle

It was clearly realized by Lavoisier that the energy for life processes, including muscular action, came from the combustion, that is, oxidation, of foodstuffs. However, by the beginning of this century it had become clear that oxidation could not be the direct source of contractile energy because muscles that were completely deprived of oxygen could nevertheless contract several hundred times. As explained briefly on p. 14, this energy comes from hydrolytic reactions—of glycogen to lactic acid, of phospho-creatine to creatine, and closest to the contractile proteins, of ATP to ADP.

Only a little ATP is present in the muscle, enough for only about eight twitches; but the rest of the chemical machinery is arranged so as to prevent this concentration from falling. If the ATP level does fall appreciably, the muscle goes into rigor, and this is exactly what happens when rigor mortis sets in after death.

The energy reserve closest to ATP is provided by phosphocreatine (PC) in conjunction with the enzyme creatine phosphotransferase (CPT).

Contraction leads to the reaction

$$\text{ATP} \xrightarrow[\substack{\text{actomyosin} \\ \text{ATP-ase}}]{} \text{ADP} + \text{P}_1$$

but the ATP is rapidly regenerated

$$\text{ADP} + \text{PC} \underset{\text{CPT}}{\xrightarrow{\;\;\;}} \text{ATP} + \text{C}$$

This transphosphorylation is reversible and the equilibrium constant (ATP)(C)/(ADP)(PC) is about 20. It follows from straightforward physical chemistry that of the ATP split, about 99 per cent is restored at the expense of PC. The ATP concentration falls appreciably only when the PC level has fallen to a small fraction of its normal value. Enough CPT is present so that this transphosphorylation proceeds at high speed during contraction itself. In consequence, no actual fall in ATP concentration has ever been found as a result of contraction unless the muscle was practically exhausted. This led some investigators to doubt whether ATP was indeed the fuel for muscular contraction. However, it has recently been found that CPT can be selectively inhibited by fluorodinitrobenzene and in these inhibited muscles, which contract more or less normally for a few twitches, ATP is indeed broken down while PC is not.

In a normal muscle the depleted stores of PC must finally be restored

to their original level during a recovery process which continues long after
contraction has ceased and which utilizes energy derived from foodstuffs.
The useful work done during this recovery process, whose details will be
described shortly, is to take in ADP and rephosphorylate it to ATP. However, when the ATP concentration rises slightly, the CPT reaction goes in
reverse and a large rise in PC concentration results from a small rise in
ATP concentration.

The main features of the recovery process are illustrated in Fig. 6–1,
which shows how glycogen is progressively dismembered in such a way as
to conserve as much as possible of its free energy.

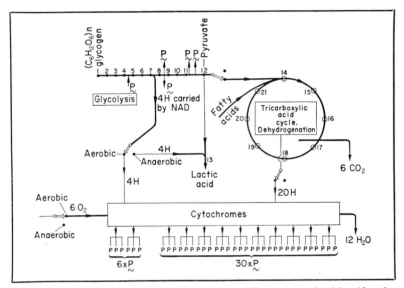

Fig. 6–1 Chemical processes in muscle. The numerals identify the
successive reactions. The quantities are appropriate for the metabolism of one
hexose unit of glycogen. The number of rephosphorylations yielded by each
stage is shown round the perimeter of the figure: each **P** presents one
rephosphorylation. The 'switches' indicate the pathway under aerobic
conditions. (From STARLING and LOVATT EVANS, 1962, *Principles of Human
Physiology* (13th ed.). Eds. DAVSON and EGGLETON. Churchill, London.
Fig. V.35.)

The first main process, consisting of twelve sequential reactions, is
called glycolysis. The glycogen is first divided into 6-carbon (hexose)
units, each of which is subsequently divided into two 3-carbon units,
ending up as pyruvic acid, $CH_3 . CO . COOH$. Apart from this, the useful
yield (per hexose unit) is three rephosphorylations, indicated by 3**P**, and
four hydrogen atoms. The hydrogen atoms are carried in combination

with special carriers such as nicotinamide-adenine dinucleotide (NAD): the transition NAD $\longrightarrow$ NADH is readily reversible. The second part of the mechanism is the tricarboxylic acid cycle (often called the TCA cycle or Krebs cycle) which accepts pyruvic acid and other substrates such as fatty acids and progressively dismembers them into hydrogen, carried in reversible combination with NAD, and carbon dioxide which is allowed to diffuse away as it is a waste product. The chemical transformation proceeds in such a way that free energy is conserved.

Most of the useful rephosphorylation occurs in the third part of the mechanism, the cytochrome chain. This is a chain of iron-containing proteins contained within the mitochondria (as are the enzymes for the TCA cycle). In the cytochrome chain, molecular oxygen is allowed to combine with hydrogen brought in as NADH, to form water. Again the free energy, or part of it, is conserved and employed to rephosphorylate ADP. Although much is known about this process of oxidative phosphorylation, the nature of the coupling at molecular level remains unclear. The iron atoms in the cytochromes (unlike those in haemoglobin) certainly undergo reversible changes between ferrous and ferric form. This change involves the movement of electrons, and it is thought that the electrons are actually transferred from one member of the chain to the next. Since there is also evidence that the cytochromes are arranged in geometrical sequence within the mitochondrion it seems that an electric current must be flowing. The process is thus analogous to what happens in a hydrogen–oxygen fuel cell—and the exact reverse of what happens during the electrolysis of water.

6.2.1   Lactic acid formation

The sequence of reactions only proceeds as shown in Fig. 6–1 if there is an abundant supply of oxygen to maintain the concentration of NADH at a low level. If the concentration of NADH rises, as a result of exercise or anoxia, it reacts with pyruvic acid to form lactic acid

$$CH_3COCOOH + NADH + H^+ \longrightarrow CH_3CH(OH)COOH + NAD^+$$

The NADH used in this reaction may be that which was produced at an earlier stage in glycolysis, so formation of lactic acid, with a small but useful yield of ATP, can proceed quite independently of oxygen, cytochrome or the TCA cycle. The formation of lactate corresponds simply to the hydrolysis of carbohydrate, and this constitutes an important energy reserve during exercise, as we shall see later. It was thought at one time that during recovery from exercise, 80 per cent of the lactate was built back into glycogen, but it is now known that this cannot occur in muscle. When oxygen is available the lactic acid is all oxidized via the TCA cycle. In an intact animal, much of the lactic acid is removed via the circulation and oxidized in other organs, notably in the heart muscle.

6.3 Heat production in muscle

We all know how exercise tends to warm us up, and the heat produced by contracting muscles plays an important part in enabling warm-blooded animals to maintain a constant body temperature. The details of the heat production by muscles have been intensively studied in the hope that they would shed light on the properties of the contractile machinery. The actual temperature rise in a single muscle twitch is very small—only a few thousandths of a degree—but the technique for measuring it, employing sensitive thermopiles, has been brought to such a high state of development through the work of A. V. Hill that the sensitivity and time-resolution exceed by a large margin what is currently possible by chemical methods.

6.3.1 Initial and recovery heat

It has been known for more than a century that heat is produced while a muscle is contracting: this heat is called the *initial heat*. One of A. V. Hill's first great contributions was to show that a roughly equal quantity of heat, the recovery heat, is produced during the period lasting many minutes afterwards when it is recovering in oxygen: this is the outward sign of the complex chemical processes that have just been described.

The time-course of the initial heat production has been studied in great detail and it has been found that when a muscle is shortening, it produces heat at a higher rate than when it is contracting isometrically. If the muscle is shortening and lifting a load, the mechanical work appears as a further extra component of the energy production. Conversely, if the active muscle is stretched, its own rate of energy production is suppressed below the isometric value. Since the energy can only come from chemical reactions, the existence of coupling between mechanical and chemical events is thus clearly demonstrated.

The basic relationship between heat and chemical change is that imposed by the first law of thermodynamics. Over any given interval of time, if only one reaction is involved:

$$\text{Heat} + \text{work produced} = n(-\Delta H) \quad \text{kcal}$$

n is the number of moles of reaction, ΔH the heat, or strictly speaking, the enthalpy change, of the reaction in kcal/mole. The sign convention is an awkward one; $(-\Delta H)$ is a positive number for an exothermic reaction.

If a muscle is completely deprived of oxygen by keeping it in pure nitrogen, and if the formation of lactic acid is prevented by applying the enzyme inhibitor iodoacetic acid, then it is thought that only one net reaction does occur, the hydrolysis of phosphocreatine. Experimentally, as Fig. 6–2 shows, the output of (heat + work) is clearly directly proportional to the breakdown of PC under many different conditions of contraction. This supports the idea that PC breakdown is the only source of energy,

4*

indicates that the *in vivo* value of ΔH is $-11\cdot1$ kcal/mole, and suggests that the variations in the time-course and amount of energy output that are seen in studies of heat production arise from variations in the rate of this single reaction.

6.3.2 *Efficiency*

As with any machine, we should like to know how efficiently muscle converts chemical energy into mechanical work. In order to do this, as

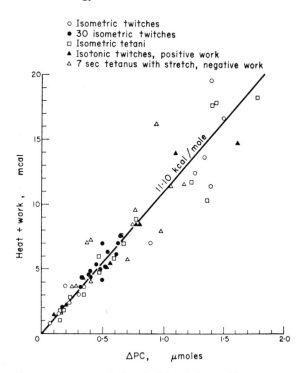

Fig. 6–2 Energy output and chemical breakdown. Heat output and work output is directly proportional to phosphocreative breakdown no matter what the type or duration of stimulation, so long as recovery is prevented by excluding oxygen and by preventing lactic acid formation with iodo acetate. From Wilkie, *J. Physiol.*, 1968, **195**, 157-83.

explained on p. 50, it is essential to know the free energy change, ΔF and not merely the enthalpy change ΔH.

For the whole process of contraction *and recovery in oxygen* it is known that $\Delta F \simeq \Delta H$ and measurements of heat production and oxygen usage agree in showing that the maximal efficiency is 20–25 per cent for frogs, toads and men. This is low compared with the efficiency of large steam turbines (35 per cent), diesel engines (40 per cent) and fuel cells (60–80 per cent). However, at present there seems to be no way of measuring the efficiency of contraction alone. Of the 75–80 per cent of the chemical energy that is wasted, we do not know what proportion is lost during the conversion of chemical energy into mechanical work, during activity; and what proportion during the much slower, purely chemical, processes of recovery.

The action of muscles in the body 7

We have seen in the preceding chapters that muscles are machines, made of protein, whose function is to convert chemical energy *directly* into mechanical work and force. A steam or petrol engine also converts chemical energy into work, but in two stages—the chemical free energy is first degraded to heat, and part of the heat is then converted into work. Direct conversion is potentially much more efficient, just as a fuel cell can be more efficient than a steam engine.

The chemical energy is derived from the hydrolysis of PC. Thus all the ingredients required for contraction are present in the resting muscle, just as in gunpowder everything is present that is needed for the explosion. Only later, during recovery, must the store of PC be regenerated, using oxygen brought from the atmosphere to oxidize food stores such as glucose and fat.

7.1 Power production by muscles

As with most other machines, the power that a muscle can produce depends on the conditions of loading. If the load is very light it will be lifted rapidly, but the power produced (force × speed) will be small. Similarly, if the load is very heavy, so that it can scarcely be lifted at all, power production will be small. However, in both these cases, chemical energy will be consumed more or less as usual. The quantitative relation between power output and force or speed was shown in Fig. 4–7, where it was demonstrated that for greatest power output, load and speed should have about one third of their maximal values. In the region of this optimum, fast mammalian and human muscles can produce about 0·2 to 0·3 horse-power per kilogram during the course of a single movement.

7.2 The action of muscles in the body

Individual muscles cannot do anything but pull; in particular they cannot push. The complex pattern of movement of our body results from two things: firstly, the anatomical arrangement by which the muscles are attached to the system of levers that constitute the skeleton; and secondly, the neurological arrangements by which different muscles are switched on and off with fine gradations of strength so that a controlled and co-ordinated movement results. Both of these arrangements are more subtle than appears at first sight; so much so that to this day complex movements remain extremely difficult to analyse fully.

The chief anatomical subtlety is that almost all the limb muscles run over two or more joints, so that when they contract, several quite different movements may be produced in several joints. The neurological subtlety is a direct consequence of this, as it is the job of the nervous system to neutralize those actions of the muscle that are *not* required, by causing contraction of other muscles that antagonize the unwanted movement. Thus, with very few and trivial exceptions (such as blinking the eye) one never finds, in an intact and healthy human being, that a single muscle contracts by itself. Almost always a whole set of muscles contract together or in sequence. To take a relatively simple example, the biceps muscle is attached at its upper end to the outer tip of the shoulder blade and at its lower end to the ulna, one of the two bones of the forearm. It is *not* attached to the humeris, the bone of the upper arm. When the biceps contracts, three movements therefore tend to occur:

1. The elbow tends to flex.
2. The forearm tends to rotate into a palm-upwards position.
3. The upper arm tends to rise away from the side of the chest.

Suppose that the movement required is (2), as would be the case in driving a screw with the forearm horizontal, then movement (1) must be neutralized by a powerful contraction of triceps (the muscle on the back of the upper arm). The occurrence of this contraction can easily be verified by feeling the triceps muscle. Also triceps is attached to the shoulder blade in such a way that its contraction tends to neutralize movement (3). In reality, the movement is even more complicated than this, and numerous other muscles participate.

7.3 The control of muscular contraction in the body

It is clear from the previous section that the control of the contracting muscles must be a very complex task. The brain and spinal cord exercise this control through the motor nerve fibres. However, each muscle fibre does not have a 'private line' from the central nervous system: each nerve fibre (axon) branches to supply a group of muscle fibres (see Fig. 7-1 (a)) and when impulses come down the axon, all the members of this group, called a *motor unit*, must contract together. The number of muscle fibres in a motor unit varies according to the fineness of control that is required. Thus in the muscles that move the eyeballs, only about ten muscle fibres are linked together, while in the biceps each motor unit contains more than a thousand muscle fibres.

7.4 Electrical recording

The activity of the muscles can be most conveniently investigated by recording the electrical changes that are associated with the passage of action potentials along the muscle fibres. One practical way of doing this, as shown in Fig. 7-1, is to stick a *concentric needle electrode* into the muscle.

An electrode of this type is easily made from a fine hypodermic needle by cementing an insulated wire down its centre, preferably with an epoxy cement such as Araldite, then grinding or scraping the protruding part of the wire away so as to leave a flat surface. This central wire is connected to the live input terminal of an amplifier via a screened cable. The needle

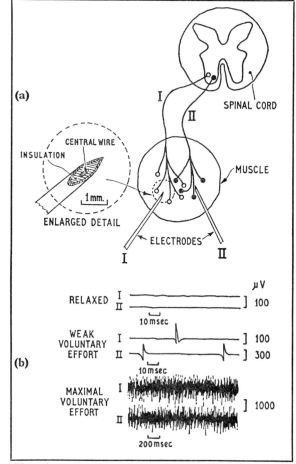

Fig. 7-1 Two electrodes, I and II, are shown in the same muscle, recording from two different motor units. (a) Concentric needle electrode. (b) Action potentials recorded simultaneously from two electrodes in human muscle. (From BUCHTHAL, 1957, *An Introduction to Electromyography.* Gyldendel, Copenhagen.)

itself is connected to the screening of the cable and thence to the earth terminal of the amplifier. A fairly sensitive gramophone amplifier can be used and the different types of signal can be identified qualitatively by the noises that are produced in the loudspeaker. For closer study of these phenomena it is preferable to record them by using a cathode ray oscilloscope and camera, when they appear as shown in Fig. 7–1 (b).

When the muscle is completely relaxed, there are no signs of electrical activity in it. When the subject makes a slight voluntary contraction, a few motor units come into action. The recording then shows a series of spikes or a series of clicks are heard. As the voluntary effort is increased, independent series of clicks arising from other motor units will be heard superimposed on the first series. Incidentally, it is possible for a trained subject to activate one particular motor unit at will. As the effort is increased still further, impulses from more and more motor units become superimposed until the individual ones can no longer be identified. The strength of contraction of the muscle as a whole is increased both by increase in the number of motor units active, and also by increasing the frequency of the impulses in the individual units.

7.5   Muscle tone

In some situations it may be essential for an animal to exert a force for a long period. Some species, for example the mussel and the oyster, have developed special smooth muscles that enable them to hold their shells closed for long periods without using up too much metabolic energy in the process. In the frog, and in a few situations in mammals—e.g. a few of the muscle fibres that move the eyeballs—there are likewise specialized striated muscle fibres. However, in general the long-continued tonic contractions are maintained by ordinary muscle fibres activated by trains of ordinary action potentials. The only specialization found is that some muscle fibres contract with a slower time course than others (see Fig. 4–4). Thus a twitch in the soleus of a cat lasts three times as long as a twitch in its peroneus longus. Probably the energy cost is similar in the two cases, but a tetanus will be maintained more economically by the slower muscle. This specialization into 'slow' and 'fast' types of muscle arises quite late during the development of the animal and is brought about by an influence, as yet not fully understood, that is transmitted down the motor nerves. This is shown by the fact that if the motor nerves to the two muscles are crossed over in the young animal, and the distal parts of the nerves are allowed to regenerate, what would have been a 'fast' muscle becomes a 'slow' one and vice versa.

There is a widespread, but incorrect, belief that all skeletal muscles are continually in a state of mild tonic contraction. However, electrical records such as those of Fig. 7–1 (b), top line, show no activity at all in a relaxed muscle. This wrong belief may have arisen by generalizing unwisely from a few particular cases; for example, that of the hind limb of a dog or cat which, because of its zig-zag arrangement, can only be kept from collapsing by tonic contraction of the extensor muscles. Animals with impairment of the central nervous system may show tone in all muscles, for example in decerebrate rigidity; and most of the disability in humans with Parkinson's disease arise from the same cause. However, in an intact animal the muscles operate in a more sensible fashion. They are active when, and only when, there is need for a mechanical force—to maintain a posture, to carry out a movement or to resist one; or by simultaneous contraction of flexors and extensors, to convert a limb into a rigid pillar.

7.6   Muscular exercise

We are all familiar with the fact that there is a limit to the intensity with which we can exercise our muscles. Beyond this limit movement becomes painful and finally impossible. The essential problem in achieving a high level of muscular activity, and thus of athletic performance, is to maintain an adequate supply of chemical energy to the contractile proteins. This energy, as we have seen, comes from two rather distinct sources:

1. Hydrolysis of PC and of glycogen to lactic acid. All the constituents are already present in the muscle, so the *rate* of the reaction can be very high, though the *total amount* of energy is limited.

2. Oxidation of carbohydrate and fat. In this case the *total amount* of energy is virtually unlimited (you could bicycle more than 2000 miles on a gallon of fat—an amount that many adults regularly carry around with them!) but the *rate* of energy supply is limited by the slowness of oxidative phosphorylation and by the complexities of supplying the active muscles with atmospheric oxygen.

The practical significance of these two energy sources is illustrated in Fig. 7–2 which shows how the power output of a man (a champion athlete in this case) falls off as the duration for which he must keep on working increases. The curve shows two distinct phases. For short durations of exercise there is a 'lump sum' of hydrolytic energy available, amounting to about 0·6 horsepower–minutes which can be generated over a long or short period as required. Superimposed on this is the energy available from oxidation (dotted line) which takes about a minute to reach its full value of about 0·4–0·5 HP but then can continue more or less indefinitely. This significance of this for athletic performance is shown by the crosses, which show the average running speed in events of various durations. Clearly, for events lasting for more than about four minutes, almost all the energy must come from oxidative sources: these permit a marathon runner to keep up for several hours a speed only slightly less than is achieved in a five kilometre race that lasts only about ten minutes.

The relative importance of hydrolytic and of oxidative metabolism is not the same in all animals—according to their way of life, many have concentrated either on one or the other. For example, the white muscles of the rabbit enable it to sprint rapidly to its burrow, while the redder muscles of the hare (redder because they contain more cytochrome and myoglobin) make possible the long endurance that the hare's existence in open country demands.

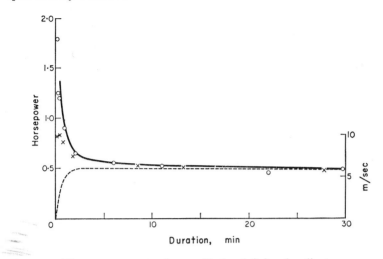

Fig. 7-2 The power output of man. Circles, left-hand ordinate, measured power output. Crosses, right-hand ordinate, running speed. The figures refer to champion athletes: trained, but normal, subjects can produce 60–80 per cent as much power. (From WILKIE, 1960, *Ergonomics*, **3**, 2.)

7.7 Effects of exercise on circulation and respiration

One of the almost immediate effects of exercise is an increase in the rate of oxidative phosphorylation which is triggered off by the slight rise in the level of ADP that occurs. Of course, this increase can only occur if extra oxygen is transported to the active muscles from the atmosphere, which in turn demands an increase in the rate at which blood is supplied to the active muscles and also an increase in the rate at which atmospheric air is processed by the lungs. The amount of oxygen stored in the muscles is small even in those red muscles which contain the haemoglobin-like protein myoglobin. Probably the function of the myoglobin is to speed up diffusion of oxygen rather than to function as a store for it.

Very soon after exercise begins, the small arteries in the active muscles dilate, thus permitting more blood to flow into the thin-walled capillaries. From these, oxygen passes rapidly by diffusion into the muscle fibres. Despite a great deal of research, it is still not certain exactly what causes this local increase in the circulation. The mechanism does not depend on the nerves that connect the blood vessels to the central nervous system: the dilatation is certainly produced by some chemical substance that the muscle fibres produce when they are active, but it is still not clear what this substance is.

If the only effect on the circulation were to reduce the resistance to blood flow in the active muscles, this would be to some extent self-defeating because the blood pressure would fall as a result and this would tend to diminish the blood flow. However, the circulation as a whole has an effective control mechanism that prevents this from happening. In the walls of part of the carotid arteries and also the arch of the aorta, there are specialized receptors which measure the blood pressure and signal it up to the brain. In response, signals are sent down to the heart which increases its rate of beating. To some extent the heart can increase its output by pumping out a small extra volume of blood each time it beats, as shown in the first part of Fig. 7–3 (b), but most of the increase in output comes from the increased rate, which increases in proportion to the severity of the exercise. The output per beat actually diminishes when the heart is beating very fast because there is so little time for it to fill with blood between one beat and the next. The cardiovascular compensation is so effective that, instead of falling, the blood pressure actually *rises* during exercise, which helps still further to maintain a rapid bloodflow to the active muscles. In Fig. 7–3 the severity of exercise has been expressed in terms of the oxygen consumed. Of course, it is only legitimate to do this if none of the energy is coming from continued net breakdown of the hydrolytic energy stores; the experimental subject must be in a 'steady state', in which the rate of oxygen intake is sufficient for the level at which work is being performed.

Each litre of oxygen that is consumed yields approximately five kilocalories of energy. Under optimal mechanical conditions, the consumption of 1 litre of oxygen per minute leads to an output of mechanical work of approximately 0·1 horsepower. Human lungs are not very effective in extracting oxygen from the atmosphere. The gas expired from them still contains at least 16 per cent oxygen—less than one quarter of the oxygen that has been breathed in has actually been absorbed. It is necessary to process about 20 litres of air in order to absorb 1 litre of oxygen. These figures remain approximately constant during exercise, so that in order to increase oxygen consumption, both the depth and the frequency of breathing must be increased considerably. The main factor responsible for this increase is not the diminution in the oxygen level in the blood but rather the rise in the level of carbon dioxide, which stimulates chemoreceptor cells in the brain. Stimulation of these receptors leads by a reflex mechanism to the required increase in depth and frequency of breathing.

7.7.1   *Recovery from exercise*

It is a fact of everyday experience that the rapid heart beats and panting respiration that accompany exercise do not return to their resting level immediately when exercise ends. The reason is that during exercise the concentrations of the various metabolites have been altered—for instance, the level of phosphocreatine has been diminished—and extra energy must be provided during recovery in order to recharge them all to their normal resting values. The extra oxygen consumed during this period is known as the 'oxygen debt' and it can amount to as much as 20 litres. It represents the energy that was made available during exercise from hydrolytic sources, but not 'paid for' at the time by oxygen consumption.

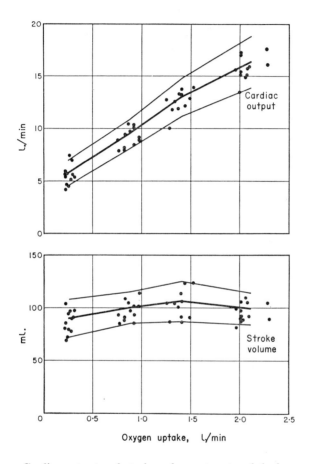

Fig. 7–3 Cardiac output and stroke volume at rest and during exercise in the supine position at 300, 600 and 900–1000 kg-metres/min in 13 healthy young men. Lines are drawn between the mean values and ± 2 × the standard deviations. Note that this exercise was taken with the subjects lying down. When standing up at rest, the stroke volume is less than is shown—about 70 ml; and it increases rapidly up to about 100 ml on even mild exercise, thereafter following roughly the curve shown. (From CARLSTEN and GRIMBY, 1966, *The Circulatory Response to Muscular Exercise in Man*, Fig. 1. Courtesy of Charles C Thomas, Springfield, Illinois.)

References

DAVIES, R. E. (1963). A molecular theory of muscle contraction. *Nature*, **199**, 1068–1074.

HUXLEY, A. F. (1957). Muscle structure and theories of contraction. *Progress in biophysics*, **7**, 255–318.

HUXLEY, A. F. and NIEDERGERKE, R. (1954). Structural changes in muscle during contraction. *Nature*, **173**, 971–977.

HUXLEY, H. E. and HANSON, J. (1954). *Ibid*, 978–987.

Further Reading

In order to obtain up-to-date information it is probably best to start with a recent review article and to work back from there. Particularly useful is the *Annual Review of Physiology*, which contained articles on muscle in 1961, 1964, 1966 and 1968.

The Abstracts—and their index—published by the Muscular Dystrophy Associations of America Inc. (1790 Broadway, New York 19) also provide a useful key to specific topics.